BIM and Urban Land Administration

BIM and Urban Land Administration

Authored by
Abbas Rajabifard, Behnam Atazadeh,
and Mohsen Kalantari

CRC Press
Taylor & Francis Group
Boca Raton London New York

CRC Press is an imprint of the
Taylor & Francis Group, an **informa** business

CRC Press
Taylor & Francis Group
6000 Broken Sound Parkway NW, Suite 300
Boca Raton, FL 33487-2742

First issued in paperback 2022

ISBN 13: 978-1-03-247546-2 (pbk)
ISBN 13: 978-1-138-49168-7 (hbk)
ISBN 13: 978-1-351-03234-6 (ebk)

DOI: 10.1201/9781351032346

Publisher's Note
The publisher has gone to great lengths to ensure the quality of this reprint but points out that some imperfections in the original copies may be apparent.

Library of Congress Cataloging-in-Publication Data

Names: Rajabifard, Abbas, author. | Atazadeh, Behnam, author. | Kalantari, Mohsen, author.
Title: BIM and urban land administration / Abbas Rajabifard, Behnam Atazadeh, and Mohsen Kalantari, authors.
Description: Boca Raton : Taylor & Francis, CRC Press, 2019. | Includes bibliographical references and index.
Identifiers: LCCN 2019001440| ISBN 9781138491687 (hardback : alk. paper) | ISBN 9781351032346 (ebook)
Subjects: LCSH: Land use, Urban--Data processing. | Land use, Urban--Government policy. | City planning--Data processing. | Building information modeling.
Classification: LCC HD108.15 .R35 2019 | DDC 333.770285--dc23
LC record available at https://lccn.loc.gov/2019001440

Visit the Taylor & Francis Web site at
http://www.taylorandfrancis.com

and the CRC Press Web site at
http://www.crcpress.com

Contents

List of Figures

List of Tables

Preface

People increasingly live in high-density urban areas which often comprise multi-story and multifunctional buildings. Multi-story buildings and urban infrastructure are needed to accommodate urban populations and provide them with various services such as transportation, telecommunications, power, and water. Urban land administration practices currently rely on 2D plans to subdivide, record, manage, and visualize the spatial dimensions of legal arrangements assigned to properties which are stratified and interlocked within multi-story developments. The efficient communication and management of legal ownership in properties of multi-story developments play an underpinning role in resolving issues among a broad community of interest holders.

This book describes the theoretical basis and practicality of connecting urban land administration practices with the 3D digital data environment of Building Information Modeling (BIM). BIM is a widely used 3D digital modeling approach in the architecture, engineering, and construction (AEC) industry. The focus of this book is to adopt a BIM-based paradigm for enhancing the communication and management of complex legal arrangements of properties in multi-story buildings which are prevalent in urban environments. BIM has emerged as a new paradigm in the AEC industry to facilitate the communication of 3D building information among actors involved in design and construction of buildings, as well as the management and operation phases. There is a strong potential in BIM to advance current land and property management systems in terms of visual communication and digital management of legal arrangements in complex urban built environments.

In current urban land administration practices, 2D paper-based plans are mainly used to delineate legal boundaries of vertically stratified private, communal, and public properties in multi-story developments. The growing dominance of these developments has recently posed significant challenges for 2D-based practices in terms of communicating and managing complex legal arrangements in urban environments. The 2D plans comprising multiple pages which are commonly used by land registries cannot be easily understood or visualized outside the domain of highly specialized professional surveyors. For the inexpert community of owners, and even land administration experts in some circumstances, there are recognized communication problems associated with spatial complexity of legal boundaries and arrangements of vertically located real properties.

The lack of an efficient and effective 3D digital solution for urban land administration limits: the ability of the resident community to visualize and understand the rights, restrictions, and responsibilities of their buildings; the ability of local government and developers to visualize multi-story developments for regulatory and registration purposes; and the ability of the government to administer a title registration system that can accommodate these increasingly complex multi-level developments. Therefore, land administration systems are seeking to adopt 3D digital models to aid in the comprehension of legal rights, restrictions, and responsibilities in multi-story developments. This book showcases the feasibility of one such model—BIM—as it offers a collaborative, 3D, digital, and semantically enriched data environment to support the integrated management of both the physical and functional aspects of buildings over their lifecycle.

BIM models provide a digital data-sharing space which is a reliable basis for making multidisciplinary decisions over various phases of the building lifecycle. From the perspective of urban land administration, the BIM environment in its current form currently provides highly detailed physical information describing buildings; however, it is not capable of managing legal information associated with the ownership of properties inside and around multi-story developments. Therefore, this book explains how the BIM environment can be extended to support the management and representation of legal ownership for properties located inside and around multi-story developments.

This book first elaborates on a range of data elements required for managing legal information in current land administration practices pertaining to the subdivision of legal arrangements within multi-story developments. It is then explained how an open data model in the BIM domain—Industry Foundation Classes (IFC)—can be extended with these legal data elements to lay the foundation for adopting a BIM-based approach in urban land administration practices. BIM models of multi-story buildings are then implemented to demonstrate the viability of using BIM for urban land administration. In addition, this book highlights the benefits and challenges of the BIM-based paradigm for urban land administration and its impacts on the governance of urban built environments. The BIM environment will play a major role in implementing future land administration systems. Therefore, this book concludes with a glance at the major directions for the future of BIM-enabled urban land administration.

This book would be of interest to many jurisdictions around the world. More specifically, it has appeal in those countries with highly dense urban areas, and megacities in the Americas, Asia, and Europe. This book is significantly relevant to countries that are using strata title systems to manage legal ownership of real properties in buildings. These countries include Australia, New Zealand, Canada, Singapore, Malaysia, Fiji, Philippines, India, and the United Arab Emirates.

This book is the first book publication connecting land administration practices with the BIM environment. Existing books mainly focus on either land administration or BIM separately. The content in this book provides a basis to underpin the intersection and integration of BIM and land administration domains, making it stand apart from other books on the subject.

We hoped to write a book that could be easily read and understood by academics, industry professionals, and students in a wide variety of disciplines such as surveying and geomatics, land administration, infrastructure engineering, urban planning, geographic information systems, and building and facility management.

Abbas Rajabifard
Behnam Atazadeh
Mohsen Kalantari

Acknowledgments

The authors sincerely acknowledge the assistance, advice, and support provided by many colleagues during the preparation of this book and in its publication. As academics and researchers, we have benefited from the international experiences and knowledge of colleagues in academic institutions, professional organizations, governments, and the private sector globally through discussions, joint research projects, and professional activities.

We acknowledge the support and work of colleagues in the Centre for Spatial Data Infrastructures and Land Administration in the Department of Infrastructure Engineering at the University of Melbourne, Australia, who helped us through the journey of writing this book. In particular, we would like to offer our special thanks to Dr Ida Jazayeri and Mr Afshin Mesbah for their input, invaluable advice, and technical knowledge. In addition, we would like to thank Dr Serene Ho for her assistance in the final stage of preparing this book.

We are also grateful for the support and work of colleagues in Land Use Victoria, State Government of Victoria, Australia; the Intergovernmental Committee on Surveying and Mapping–ICSM; the City of Melbourne; buildingSMART Australasia and Reeds Consulting. Their in-depth experience and knowledge helped us to accomplish this journey.

The staff at Taylor & Francis Group/CRC Press have been very supportive and understanding during the inevitable challenges in writing this book. We would like to express our sincere gratitude for their commitment and professionalism, and many thanks in particular to Ms Irma Britton.

Abbas Rajabifard
Behnam Atazadeh
Mohsen Kalantari

Authors

Abbas Rajabifard is a professor in Geomatics Engineering and Director of the Centre for Spatial Data Infrastructures & Land Administration (CSDILA) at The University of Melbourne. He is also Chair of the United Nations Global Geospatial Information Management (UN-GGIM) Academic Network, which is a research and training strategic arm of UN-GGIM. Prof. Rajabifard has an active research interest in the areas of urban land administration, building information modeling, spatial enablement, 3D cadastre, and urban data analytics to support SDGs, resilience, and smart cities. He has spent his career researching, developing, applying, and teaching SDI, land administration, and spatial enablement to deliver benefits to governments, industry and wider society. He has widely consulted and published in these areas.

Behnam Atazadeh is a research fellow in urban land administration at the Centre for Spatial Data Infrastructures and Land Administration, the University of Melbourne. He has research experience in using 3D building information environments for advanced land administration in urban areas. Dr Atazadeh is a committee member of Standards Australia where he is involved in the development of building information modeling standards. He has published several articles in reputable scientific journals and conferences in the domains of surveying and spatial information sciences.

Mohsen Kalantari is a senior lecturer in Geomatics Engineering and Associate Director at the Centre for SDIs and Land Administration at The University of Melbourne. He has experience in developing successful research grant applications, engaging with the industry and profession, building strong research teams, and conducting research in Australia and overseas. His areas of research include land administration, cadastre, spatial data infrastructures, 3D GIS, and building information modeling. Beyond his contributions to academia, Dr Kalantari has had a demonstrable contribution to industry; most notably, he is a co-founder of a start-up company on the automatic conversion of outputs of laser scanners to BIM and 3D GIS.

Introduction

This book describes the theoretical basis and practicality of connecting urban land administration practices with the 3D digital data environment of Building Information Modeling (BIM). BIM is a widely used 3D digital modeling approach in the architecture, engineering, and construction industry. The focus of this book is to adopt a BIM-based paradigm for enhancing the communication and management of complex legal arrangements of properties in multi-story buildings which are prevalent in urban environments.

This book first elaborates on a range of data elements required for managing legal information in current land administration practices pertaining to the subdivision of legal arrangements within multi-story developments. It is then explained how an open data model in the BIM domain—Industry Foundation Classes (IFC)—can be extended with these legal data elements to lay the foundation for adopting a BIM-based approach in urban land administration practices. BIM models of multi-story buildings are then implemented to demonstrate the viability of using BIM for urban land administration. In addition, this book highlights the benefits and challenges of the BIM-based paradigm for urban land administration and its impact on the governance of urban environments.

This book comprises seven chapters:

Chapter 1, *Introduction—Managing Legal and Physical Complexities of Urban Environments,* provides a basic introduction to capturing legal and physical aspects of urban built environments. This chapter also introduces current 2D-based practices adopted for legal partitioning of multi-story buildings, and highlights spatial problems associated with these practices in terms of communicating and managing legal ownership of multi-story properties. To address these problems, 3D legal and physical models of buildings are suggested in this chapter. In addition, this chapter introduces BIM, and describes the benefits and challenges of this 3D digital approach in the AEC industry. Moreover, the chapter highlights the potential of BIM for adding value in upgrading current urban land administration systems into a 3D digital, intelligent, and dynamic environment.

Chapter 2, *Urban Land Administration—A Digital Paradigm,* introduces the general concept of urban land administration as recording, managing, and communicating legal information in complex urban areas. In practice, each jurisdiction considers its own guidelines for implementing land administration systems. The common aspect of these practices is their reliance on devising fragmented 2D-based representation methods for managing complex RRR spaces prevailing in urban areas. Enhancing current 2D-based systems to 3D digital environments requires a good understanding of data requirements in urban land administration. Therefore, this chapter provides a thorough explanation of urban land administration processes in Victoria, Australia, which helps to identify the main data requirements. Finally, this chapter describes various 3D data models which have recently been adopted at an international level for realizing digital paradigms in urban land administration.

Chapter 3, *Fundamentals of the BIM Environment—Opportunities for Land Administration*, provides a general introduction to the fundamental concepts behind the paradigm of BIM and elaborates its importance in the land administration domain. Moreover, open BIM standards and relevant IFC entities for urban land administration are described. A comparison between the IFC standard and other 3D spatial data models, particularly the CityGML standard, will also be presented in this chapter to highlight the capabilities of the IFC standard in digitalizing urban land administration practices. In the last section of this chapter, an overview of relevant BIM projects in the land administration domain will be presented.

Chapter 4, *BIM-Enabled 3D Digital Urban Land Administration—New Era in Managing Cities*, begins by first describing different alternatives for extending the IFC standard and the reasons for choosing the most appropriate extension method. Next, the major parts of the adopted extension method are described in detail. This is followed by an explanation of the details for modeling each legal data element within the extended version of the IFC standard. It should be noted that since legal data elements are highly dependent on the jurisdictional setting, the extension of the IFC standard presented in this chapter is based on those elements identified in the context of the Victorian jurisdiction. However, the chapter concludes by providing general, jurisdiction-independent guidelines for realizing a BIM-based approach for urban land administration.

In Chapter 5, *Implementing BIM Models—From Simple to Complex Urban Developments*, to demonstrate the feasibility of a BIM-driven approach for urban land administration, prototype BIM models are implemented for simple and complex developments located in an urban environment. These implemented prototype models showcase the practicality of recording, managing, and visualizing legal and physical dimensions of multi-story developments within the 3D digital data environment of BIM. The chapter considers two approaches for constructing BIM models: 3D design and 3D surveying. Most existing BIM models are constructed during the design phase of a building, and they are not updated after the construction phase of the building. Therefore, in this chapter, various 3D surveying methods for creating as-built BIM models are also presented.

Chapter 6, *Impacts of BIM on Urban Land Administration*, highlights the benefits of realizing a BIM-based paradigm, which is predicated on the proposed IFC extension, in current practices of urban land administration. Adoption and implementation of a BIM-based approach in the land administration domain will enable an accurate, authoritative, and unambiguous recording and representation of 3D digital land and property information, which is fundamental to ensuring security, comprehension, and ongoing commoditization of complex vertical legal arrangements common in urban environments. In addition, the technical challenges and organizational barriers in adopting a BIM-based digital urban land administration are described in this chapter. The technical issues can be overcome by finding the appropriate technical solutions. However, the existing institutional constraints, which support and reinforce the use of 2D-based land administration practices and processes, are the major challenges faced in adopting the right roadmap for 3D digital urban land administration.

Chapter 7, *Prospect of BIM in the Land Administration Domain—Technical Aspects,* enlightens the reader on the future of BIM in the land administration domain. This includes leveraging a BIM environment for managing the lifecycle of processes defined in urban land administration; expanding a BIM environment for managing the legal ownership of properties in large infrastructure projects; validating the integrity of 3D spatial data acquired from BIM; developing 3D reasoning operators to go beyond the visualization of multi-story properties; extending the IFC standard to support urban land administration in any jurisdiction; using a BIM-driven approach for measuring volumetric quantities associated with urban properties; and automating the generalization and simplification of BIM models.

1

Managing Legal and Physical Complexities of Urban Environments

1.1 Land Administration Systems and Urban Complexity

Land administration systems provide an overarching framework to design and implement effective policies and strategies associated with the planning and development of land in line with the goals of sustainable development (Williamson et al. 2010). These systems adopt an organized approach to addressing issues associated with land and property assets by considering social justice, economic prosperity, and environmental sustainability. Land administration systems are critically important for developing a national infrastructure for any country around the world. The four underpinning functions of a modern land administration system are:

- Land tenure: This refers to procedures defining and securing legal rights and boundaries associated with land and properties.
- Land value: This function specifies the assessment processes for taxation and valuation of land and properties.
- Land use: This refers to the procedures controlling the usage of land in line with planning policies and regulations developed at national, state, and local levels.
- Land development: This includes the processes for designing, planning, and building new physical structures.

In a modern society, land is a broad concept comprising the land itself, physical objects and cognitive spaces within underground and aboveground areas, physiographical features (such as waterbodies and geological substratum), and airspaces. This signifies the fact that modern land administration systems should develop and implement effective solutions for spatial management and subdivision of an entire jurisdiction in three dimensions. Three-dimensional (3D) spatial extents of rights, restrictions, and responsibilities (RRR) associated with land are currently recorded and stored using cadastral maps and subdivision plans. As outlined by the International Land

Administration Domain Model (LADM) standard, "a right may provide a formal or informal entitlement to own or do something, a restriction is a formal or informal obligation to refrain from doing something, and a responsibility is a formal or informal obligation to do something."

Throughout the last few decades, rapid urbanization and population growth have meant an unprecedented pressure on the development and use of land in cities around the world. Due to the limited areas of land in urban environments, there has been a prolific growth of built assets within these environments in the vertical dimension. Examples of these assets are multi-story, high-density, and multi-functional building developments, shopping centers, passageways above and below streets, gas and water pipelines, electricity and communication cables, underground carparks, and tunnels.

We are entering an era where more people will be living or working in multi-owned urban developments. Owners of apartments are liable to contribute to the maintenance of the apartments and common areas. Buyers or renters are given documents about what they buy or rent, or the opportunity to obtain them. This indicates that multi-story properties in urban areas are not only becoming a significant component of the land market and economy in many nations, but also generating a rising tide of social, economic, and environmental issues impeding the move towards sustainable development.

Urban land administration mainly refers to processes associated with managing legal ownership of properties in multi-story developments which are the most prominent components of an urban built environment. These developments typically comprise a wide range of RRR spaces associated with privately, commonly, and publicly owned properties. For instance, a private property can be considered as an apartment unit and its carpark, while a common property can include various common areas spanning throughout the building (e.g. corridors, lobbies, elevators, and gym).

The requirements for recording RRR information in each jurisdiction depend on the guidelines and procedures developed specifically for that jurisdiction. Therefore, each jurisdiction adopts its own unique approach for implementing land administration systems in urban areas. However, urban land administrations practices among all jurisdictions commonly rely on 2D-based methods of partitioning the spatial extent of RRR spaces within multi-owned developments.

1.2 Legal Partitioning of Multi-Story Buildings: 2D-Based Approaches

Building subdivision practices currently use 2D analogue or digital plans for delineating the legal limits of properties in multi-story buildings. These plans typically include multiple pages, each of which provides a projection of 3D ownership spaces from a specific viewpoint. These projections typically include both vertical and horizontal representations of 3D ownership spaces (see Figure 1.1). As indicated in Figure 1.2, in some cases, a volumetric representation of an ownership space from an isometric viewpoint is delineated in these 2D plans (such as in Queensland, Australia).

2D-based representation methods are quite efficient in multi-story buildings with simple structural forms. However, the efficiency of 2D plans is questionable in buildings with higher levels of spatial and functional complexities. Spatial complexity refers to architectural and structural elements defining the internal and external shape of a building in its various parts. Functional complexity indicates the usage and access of building facilities differentiated among various groups of a building's inhabitants. For example, a multi-purpose building could have various access requirements to support retail, commercial, and residential functions. Therefore, there are a number of communication challenges in delineating legal boundaries of properties when 2D-based methods of legal partitioning are used in complex buildings. These challenges include:

- Difficulties in integrated representation of boundaries of ownership spaces with irregular forms: The complex reality of ownership spaces with interweaving and irregular shapes may not be successfully recorded and managed via 2D paper-based drawings, or even 2D digital plans, and this forms the first technical challenge associated with delineating boundaries (see Figure 1.3).
- Use of numerous 2D floorplans and cross-section diagrams to fully represent all ownership spaces in high-rise buildings: A large number of 2D diagrams results in difficulties finding the right floorplan and cross-section diagrams that, when viewed together, represents the horizontal and vertical boundaries for a specific private ownership space (Jazayeri, Rajabifard and Kalantari 2014). Consequently, interpretation of ownership boundaries for each apartment unit in high-rise developments is a time-consuming task. Additionally, although 2D subdivision plans can be functional to some extent, this 2D-based method of representation may be computationally extensive, as multiple pages of 2D diagrams are used to describe RRR information associated with superimposed properties in high-rise buildings. As an example, over 50 pages of 2D plan drawings are required to effectively present the overall parcel, as well as individual property boundaries, for a 40-story apartment building in the City of Melbourne (see Figure 1.4).
- Problems in accurate understanding of textual information: 2D subdivision plans usually include textual notations (see Figure 1.5) which are difficult for nonspecialized owners, who have limited background in understanding subdivision plans, to read and interpret

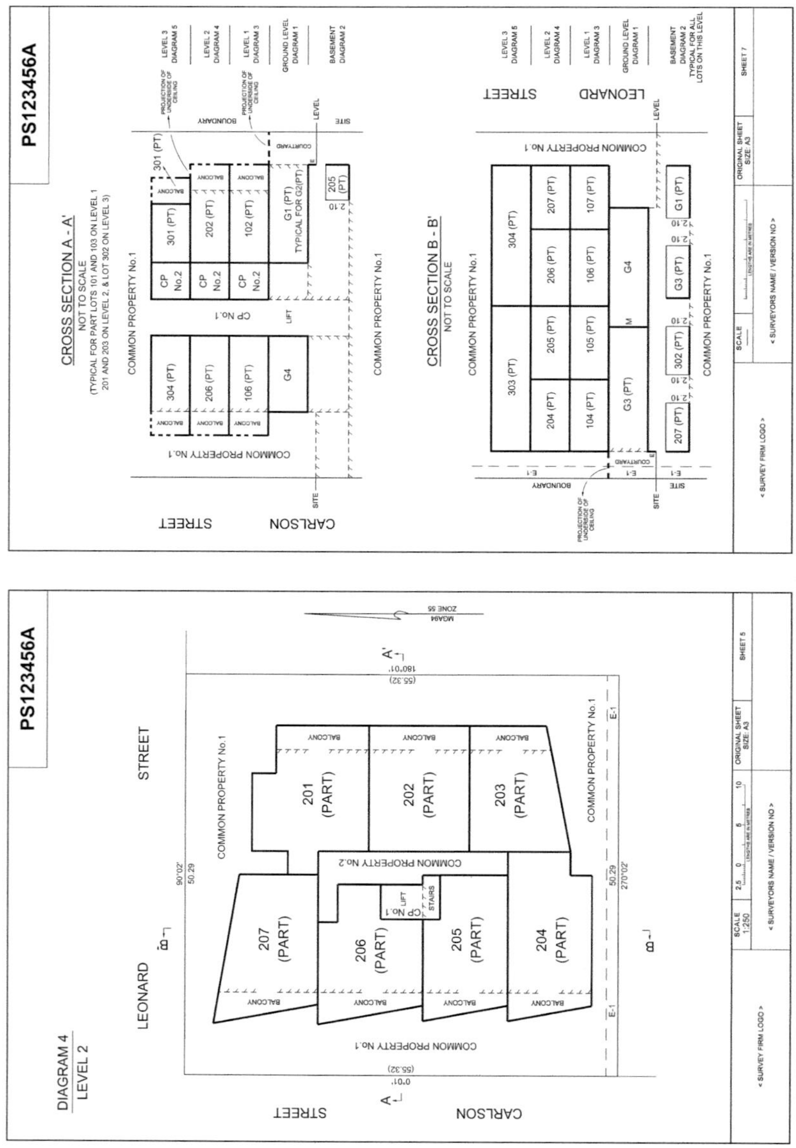

FIGURE 1.1
An example of floorplan and cross-section views of ownership spaces. (From Victoria State Government 2019.)

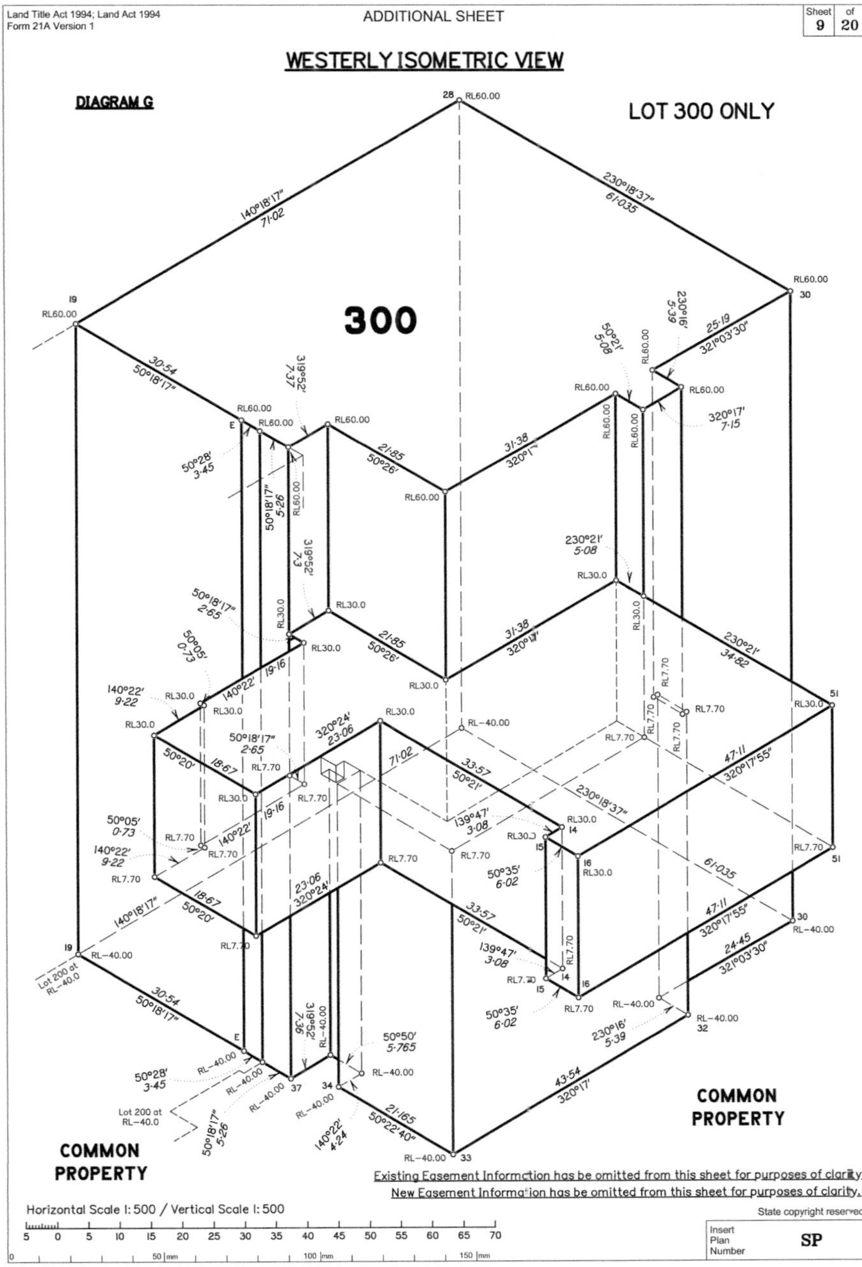

FIGURE 1.2
An example of volumetric representation of ownership space on a 2D plan. (From Land Title Act 1994.)

FIGURE 1.3
A future building structure with a complex architectural shape (Arbre Blanc tower). (Courtesy of Sou Fujimoto, Oxo, Nicolas Laisné, and Dimitri Roussel.)

accurately (Shojaei et al. 2014). This textual information typically includes notation regarding spatial location of legal boundaries in reference to the physical structure. Moreover, it can also provide information on what parts of the building belong to the communal area. Accurate understanding of these textual descriptions could be problematic for non-technical people.

- Representation of 3D RRR spaces spanning several stories of a building: While multiple 2D plans can illustrate ownership boundaries for properties restricted to one level, implementing these plans to communicate the ownership boundaries of properties encompassing parts of several levels is a very challenging task. Many multi-story buildings include commonly owned and managed properties (e.g. corridors, stairwells, gyms, etc.). These common property areas typically pass through several building levels and it is a very challenging task to map the boundaries of these areas into 2D floorplans and cross-section diagrams. This challenge is more extensive in comparison with the aforementioned challenges. In Figure 1.6, a common property area, illustrating corridors, stairs, and lobbies, of a multi-story building is shown in red to provide an example of this challenge.

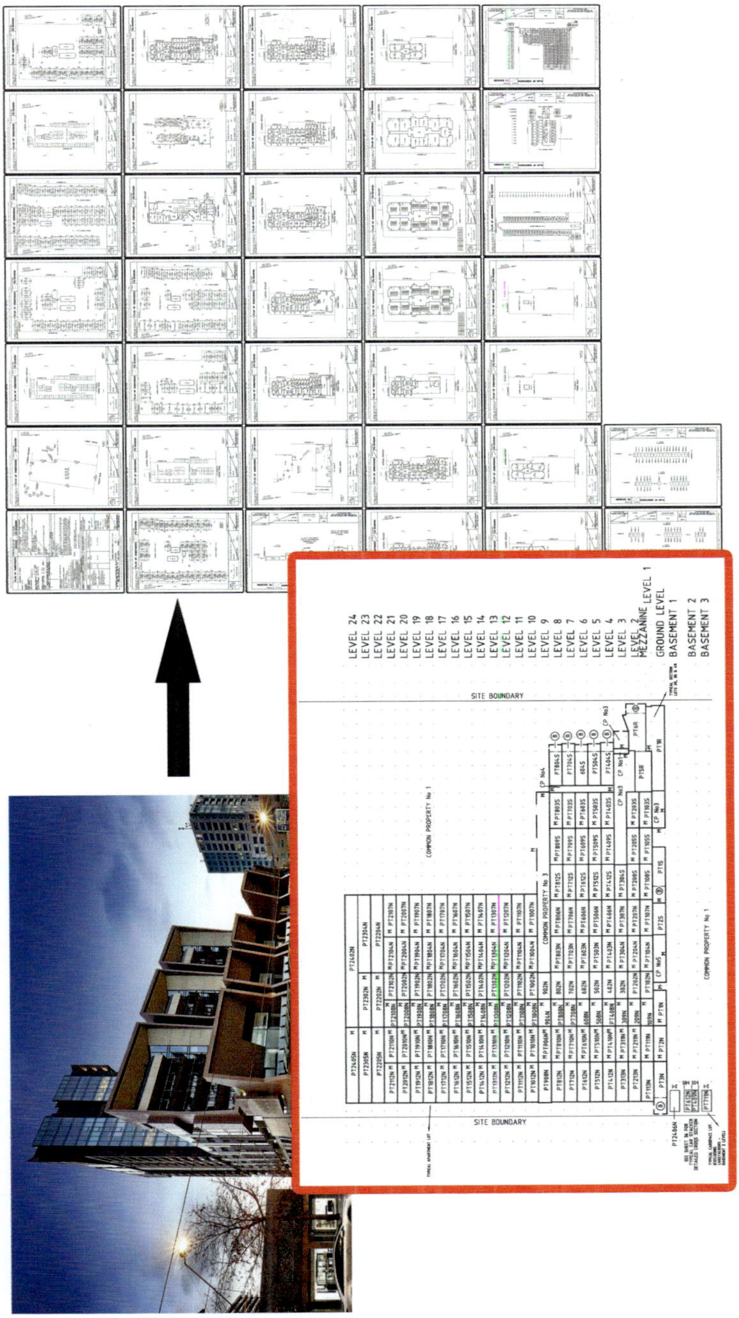

FIGURE 1.4
Multiple pages of plans for subdividing ownership spaces in high-rise buildings. (Adapted from Rajabifard et al. 2014.)

PLAN OF SUBDIVISION	EDITION 1	**PS123456B**

LOCATION OF LAND

PARISH: MELBOURNE SOUTH
TOWNSHIP:
SECTION: 2
CROWN ALLOTMENT: 34 (PART)
CROWN PORTION:

TITLE REFERENCE: VOL.80645 FOL. 649,

LAST PLAN REFERENCE: LP211551 (LOT 65)

POSTAL ADDRESS: 5 LEONARD STREET
(at time of subdivision) SOUTH MELBOURNE 3051

MGA CO-ORDINATES: E: 395705 ZONE: 55
(of approx centre of land in plan) N: 5562314 GDA 94

VESTING OF ROADS AND/OR RESERVES

IDENTIFIER	COUNCIL/BODY/PERSON
NIL	NIL

NOTATIONS

DEPTH LIMITATION: NIL

SURVEY:
This plan is/~~is not~~ based on survey.
This survey has been connected to permanent marks No(s).
In Proclaimed Survey Area No.

LOTS IN THIS PLAN MAY BE AFFECTED BY ONE OR MORE OWNERS CORPORATIONS
FOR DETAILS OF ANY OWNERS CORPORATIONS INCLUDING PURPOSE, RESPONSIBILITY,ENTITLEMENT & LIABILITY SEE OWNERS CORPORATION SEARCH REPORT, OWNERS CORPORATION ADDITIONAL INFORMATION AND IF APPLICABLE, OWNERS CORPORATION RULES

STAGING:
This is/~~is not~~ a staged subdivision.
Planning Permit No.

NOTATIONS

BOUNDARIES DEFINED BY BUILDINGS ARE SHOWN BY THICK CONTINUOUS LINES

LOCATION OF BOUNDARIES DEFINED BY BUILDINGS:
MEDIAN: BOUNDARIES MARKED 'M'
EXTERIOR FACE: BOUNDARIES MARKED 'E'
INTERIOR FACE: ALL OTHER BOUNDARIES

COMMON PROPERTY No.1 IS ALL THE LAND IN THE PLAN EXCEPT THE LOTS AND COMMON PROPERTY No.2 AND INCLUDES THE STRUCTURE OF WALLS FLOORS AND CEILINGS THAT DEFINE BOUNDARIES, EXCEPT THOSE MARKED 'M'.

ALL INTERNAL COLUMNS, SERVICE DUCTS, PIPE SHAFTS AND CABLE DUCTS WITHIN THE BUILDING ARE DEEMED TO BE PART OF COMMON PROPERTY No.1. THE POSITIONS OF THESE COLUMNS, SERVICE DUCTS, PIPE SHAFTS AND CABLE DUCTS MAY NOT HAVE BEEN SHOWN ON THE DIAGRAMS CONTAINED HEREIN.

CP = COMMON PROPERTY
PT = PART
ᚁᚁᚁᚁ INDICATES THE LOCATION OF STRUCTURE (NON BOUNDARY)

EASEMENT INFORMATION

LEGEND: A - Appurtenant Easement E - Encumbering Easement R - Encumbering Easement (Road)

EASEMENTS AND RIGHTS IMPLIED BY SECTION 12(2) OF THE SUBDIVISION ACT 1988 APPLY TO THE WHOLE OF THE LAND IN THIS PLAN

Easement Reference	Purpose	Width (Metres)	Origin	Land Benefited/In Favour Of
E-1	DRAINAGE	1.50	THIS PLAN	UPPER EAST WATER CORPORATION

< SURVEY FIRM LOGO >	SURVEYORS FILE REF:	ORIGINAL SHEET SIZE: A3	SHEET 1 OF 7 SHEETS
	< SURVEYORS NAME / VERSION NO >		

FIGURE 1.5
Textual notations in the fact sheet of a subdivision plan. (From Victoria State Government 2019.)

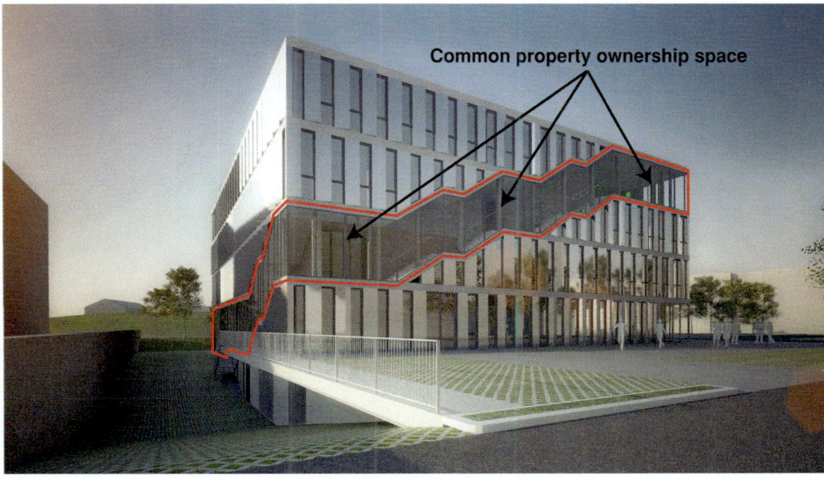

FIGURE 1.6
An indicative example of a complicated common property area.

- Ambiguities in interpretation of legal boundaries defined by building elements: Understanding of some legal boundaries, which are defined by referencing building elements, could be problematic for inexpert stakeholders. For example, in Victoria, Australia, legal boundaries inside buildings are mainly defined by using one of three spatial relationships with its corresponding building element: the boundary touches either the interior or exterior face of the building element, or it corresponds with the (virtual) median line of the building element (Libbis 2015). The referenced building elements typically include walls, windows, ceilings, slabs, and doors. These legal boundaries are often denoted by different types of lines in subdivision plans. As shown in Figure 1.7, some boundaries of LOT 1 are defined by referencing the interior face of the wall, door, and window. These physical structures are not shown in subdivision plans. This abstraction could potentially lead to a misunderstanding of these boundaries by a non-expert user interpreting what the lines mean, and hence he/she could easily misunderstand the location of the boundary and RRR information. For instance, a non-specialist owner of LOT 1 may misinterpret interior boundaries as exterior ones. Therefore, he/she may think that the full ownership and responsibility of building elements is entitled to him/her, which is not correct in this case.
- Static content and no support for querying legal boundaries: The data environment of subdivision plans is analogue and static. Therefore, it is impossible to view the boundaries of ownership spaces in an interactive mode (Döner et al. 2010; Stoter et al. 2004). This drawback

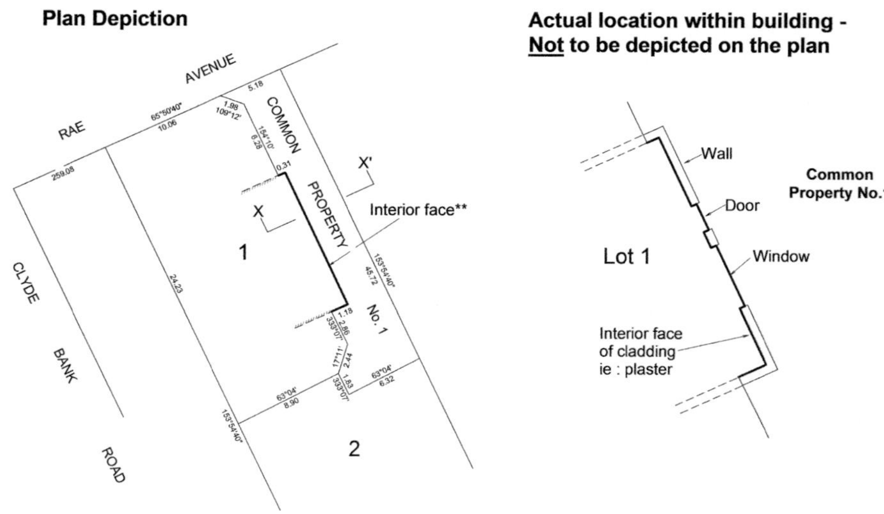

FIGURE 1.7
Boundaries referencing physical structures.

may affect the interpretation and understanding of legal boundaries in complex situations. Moreover, an analogue data environment does not provide the capability to query, find, and measure legal boundaries (Acharya 2011).

To overcome the aforementioned challenges, there has been an increasing trend towards leveraging 3D digital building models in the land administration domain. Some of these models are comprised solely of pure geometric information, such as shape, façade, and texture, about building elements, while others also comprise semantic information which would enable a wide range of 3D building information analysis (Kolbe 2009).

1.3 3D Digital Models

Recent advancements in computer graphics and 3D modeling capabilities and the new demands for mapping spatially complex situations has led to the adoption and use of 3D digital data environments in various industry sectors. Currently, 3D digital models are being used in a wide range of applications, such as urban planning, architectural design, disaster management, infrastructure, and facility services. 3D digital data environments provide more intuitive and understandable representations of objects, and the end user can interact and view 3D digital models from various viewpoints within these environments. 3D digital models were initially developed for designing new products. For instance, architects have been using 3D digital models for creating new building designs. More recently, real-world spatial objects were also represented as 3D digital models. One example is the use of drones to capture images from an urban area, which are subsequently used to create 3D digital models of existing built assets.

Relevant to modeling urban built environments, various types of 3D digital models have been created and used in industrial and governmental organizations. These models play an underpinning role in digitalizing management and operations of modern cities. In essence, urban built environments are spatially complex, and there are numerous types of structures intertwined with each other within these environments.

Creating a fully detailed model of urban objects is not always required. Therefore, 3D digital models typically provide a generalized representation of urban objects. This generalization is dependent on the expected outcome from a 3D digital model in order to use it for a specific application. For instance, urban planners might only be interested in the external 3D shape of a building for modeling shadows cast in its neighborhood, while 3D modeling of structures and spaces inside the same building could be useful for indoor wayfinding and navigation purposes.

There are three important stages in developing and implementing 3D digital models, namely, 3D data sourcing, 3D data modeling, and 3D data visualization and analysis. There are different acquisition approaches to sourcing and recording 3D digital data. These include: processing aerial imageries and point clouds, architectural drawings for buildings and infrastructure, extrusion of 2D footprints of objects, constructing 3D models from 2D architectural plans, rule-based procedural modeling and shape grammars, and crowdsourcing approaches. Once 3D digital data is acquired, it will be stored in either a file or a database. The basis of structuring and storing 3D digital data is dependent on the adopted 3D data model. 3D data models provide a logical arrangement of spatial objects and their relationships in a 3D digital data environment. Well-known examples of 3D data models are open CityGML (Groger et al. 2012) and Industry Foundation Classes (IFC) (ISO16739 2013), standards which are in use in the geospatial and building industries, respectively. The final stage is to visualize and analyze 3D digital data for a specific use.

A wide range of 3D visualization techniques have been developed to better interact and inspect 3D digital models. Some examples include various tools for navigating around and inside 3D digital models (such as pan, zoom, and rotation), generating plans and section views of 3D models, hiding or showing specific objects, and viewing models from various perspectives. Although visualization of 3D digital data is currently very sophisticated, 3D data analysis is still in its infancy stages, and there are limited approaches developed for analyzing 3D digital models, particularly 3D spatial analysis associated with indoor built environments, due to their architectural and structural complexities. Some examples of 3D analysis used for outdoor urban areas are modeling noise propagation, computing solar irradiation on building roofs, visibility analysis, and outdoor thermal comfort analysis.

For the purpose of urban land administration, 3D digital data environments should be able to support two types of 3D spatial data, namely 3D legal objects and 3D physical objects, as well as logical interdependencies between them (Rajabifard, Atazadeh, and Kalantari 2018). A 3D legal object refers to a cognitively defined volume of space associated with a specific legal interest. For instance, the volumetric space of an apartment unit can be considered as one 3D legal object. A 3D physical object is defined as those objects which exist in the real world and are physically tangible. Examples are structural and architectural components such as slabs, walls, doors, and so on.

Boundaries of a legal object are considered as the important aspect in specifying spatial dimensions of these objects. Land administration practices typically rely on rigorous mechanisms for delineating boundaries of legal objects. Additionally, semantic attributes associated with each legal object are also important in urban land administration. For example, an attribute can be used to specify whether the ownership right associated with a legal object is a private, communal, or public one. These factors imply that the main purpose of urban land administration practices is to define spatial and semantic information associated with legal objects. However, physical objects are also required in these practices as auxiliary elements to facilitate the communication of spatial arrangements and boundaries of legal objects. For instance, the spatial location of boundaries for a legal object (e.g. a privately owned unit) can be defined through the interior face of walls (see Figure 1.7). In this case, if both wall and legal object are represented in a 3D digital data environment, it would be much easier for a lay user (e.g. the unit owner) to understand the location of her/his boundaries adjacent to other property units.

1.4 3D Legal Models of Buildings

Most efforts in digitalizing urban land administration practices mainly focused on the development of 3D legal models for buildings, as shown in various research papers published in 3D cadastre workshops commenced from 2001. One reason for the emergence of 3D legal models is that since current urban land administration practices represent 2D-based projections of only legal spaces, 3D legal models attempted to address such fragmented representations of volumetric legal spaces (see Figure 1.8). Another reason is

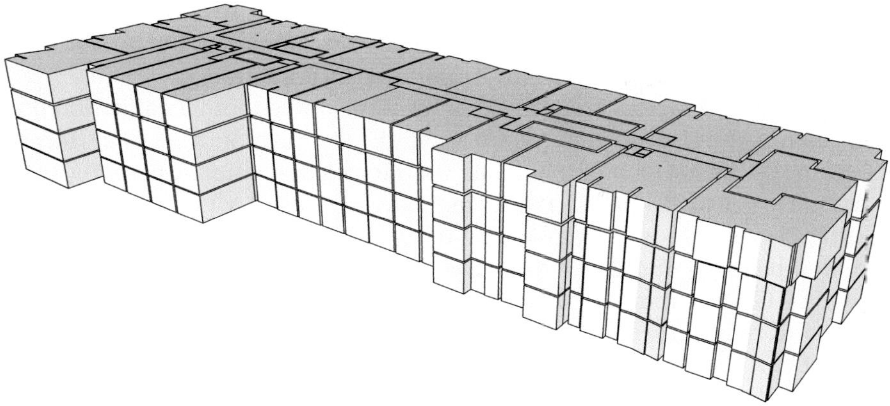

FIGURE 1.8
3D legal spaces of property units within a multi-story building.

that some jurisdictions consider legal objects completely independent from the physical objects, and only legal representation of buildings is adequate for their purpose. The vast majority of efforts in constructing 3D legal models of buildings can be observed on either developing data models or implementing prototype visualization systems.

For data modeling purpose, two 3D legal models, namely Land Administration Domain Model (LADM) and ePlan, have been considered as the most prominent ones. LADM is a widely accepted international standard for information sharing and communication in the field of land administration (ISO19152 2012). This standard provides the general concept of "Spatial Unit" for modeling 2D land parcels and 3D legal spaces. In the context of the urban built environment, two specific types of spatial units for modeling legal spaces are developed in the LADM standard. The first one is the "Legal Space Building Unit" which is specifically used for modeling and registering legal spaces inside and around buildings. The second concept is the "Legal Space Utility Network" which defines legal spaces associated with utility networks. This type of spatial unit is useful for imposing restrictions in order to protect legal rights assigned to utility networks within urban developments. The legal spaces associated with building units and utility networks do not essentially coincide with their corresponding physical spaces.

ePlan is another legal data model adopted nationally in Australia, New Zealand, and Singapore (ICSM 2010). This data model also provides the capability to model legal spaces of buildings, which are known as a "Volumetric Lot." A volumetric lot is a general concept adopted in the ePlan data model for constructing 3D legal models.

There has been a wide range of visualization prototypes developed for representing 3D legal spaces. Some of these prototypes were implemented according to the specific requirements of a particular jurisdiction (García,

Soriano, and Martín-Varés 2011; Vandysheva et al. 2012). In addition, other 3D visualization prototypes were implemented to demonstrate the feasibility of LADM and ePlan data models for urban land administration.

The important factors considered in the development of visualization prototypes are the data format, platform, and visualization tool. Data format refers to the format of 3D digital data used for visualization. Currently, there is a wide range of 3D digital data formats such as FBX, KML, OBJ, COLLADA, VRML, 3DS and so on. The platform used for visualization could be a desktop, web-based, or mobile environment. The tools used for visualization rely on CAD, GIS, or their own customized environment. Understanding the needs and requirements of end users plays a fundamental role in the development of a usable 3D visualization system for urban land administration. There is a broad range of end users in urban land administration, such as local governments, land surveyors, land registries, lawyers, owners, facility managers, and utility companies.

Shojaei (2014) identified a comprehensive set of requirements for visualizing 3D digital data in the context of urban land administration. Important examples of these requirements include handling massive 3D data, representing query results, interactivity, underground views, cross-section views, distance measurements, transparency, levels of detail, labeling and coloring 3D legal objects, and controlling 3D legal objects and layers. A good understanding and documentation of requirements would help property visualization systems to better represent and communicate 3D legal information. In particular, these include understanding ownership boundaries, locating a specific legal space, identifying spatial relationships between with two adjacent legal spaces, and clear differentiation of a privately owned property from common property inside multi-owned building developments. Visualization requirements often refer to the importance of physical objects in facilitating the understanding of legal spaces. Therefore, 3D physical models of buildings will be introduced in the next section.

1.5 3D Physical Models of Buildings

3D physical models have been developed in construction or geospatial industries to either design new buildings or capture the physical reality of existing buildings. Over the last decades, a wide range of 3D building models with various levels of detail have been created and used for various engineering and planning applications. The primary purpose of developing 3D physical models of buildings was not for urban land administration. However, they can be extended with legal data elements to support 3D-enabled urban land administration.

Depending on the use case, 3D physical models can vary from simple cubical representation of buildings to highly detailed geometric and semantic modeling of all elements inside buildings. Early developments in 3D physical models of buildings mainly focused on the geometric representation of buildings for the purpose of 3D visualization. However, 3D geometric representations of buildings were not adequate for many engineering computations and planning analyses. Therefore, semantic information about building elements was incorporated into 3D physical models of buildings. This made 3D physical models more intelligent, and provided the capability to surpass visualization and enable spatial computations associated with 3D physical data. 3D building models were developed for exchanging 3D spatial information. These 3D models can be grouped into three main categories: purely geometric models, geometric models with limited semantic information, and models comprising comprehensive geometric and semantic data structure.

In the context of urban land administration, models from the last category could be considered as the most intelligent ones, which provide the capability to map physical complexities inside urban developments. CityGML is the most prominent example used for modeling both geometric and semantic aspects of buildings in 3D digital environments. This 3D physical model provides the ability to represent a building in various levels of detail (LoD). In the coarsest LoD, buildings in CityGML can be represented as extruded blocks, while the finest LoD provides the geometric representation of interior structures and indoor spaces of complex buildings (see Figure 1.9). In addition, semantic entities associated with various building components such as doors, windows, walls, ceilings, and columns are defined by the finest LoD. The semantic relationships between building components are also supported in CityGML. For instance, there are bounding relationships between an indoor space and the walls bounding that space. Another comprehensive 3D physical model of buildings is known as Building Information Model (BIM), which is the major focus of this book. BIM provides a very rich and intelligent digital infrastructure for managing 3D building components over the entire lifecycle of buildings.

FIGURE 1.9
Various LoDs in CityGML. (Adapted from Biljecki, Ledoux, and Stoter 2016.)

1.6 Building Information Models

Similar to the demand for 3D digital data in urban land administration, the Architecture, Engineering, and Construction (AEC) disciplines taking part in the building development process have also been faced with the limitations of 2D-based designs in terms of communicating complex architectural and structural information associated with multi-story buildings (Eastman et al. 2011).

As a result, Building Information Modeling (BIM) has emerged as a new paradigm to facilitate collaboration among different AEC actors (Isikdag 2015). As stated in NBIMS (2012), BIM provides a 3D digital and shared data environment encompassing both physical and functional characteristics of built assets, underpinning a reliable basis for decision making over the lifecycle of these assets. In comparison with a 2D-based means of information sharing, BIM significantly improves productivity and reduces cost and effort for design changes in the development phases of complex and multi-story building structures (Arayici et al. 2011; Barlish and Sullivan 2012).

The "BIM" acronym is generally used to refer to two distinct concepts, namely, process and product (see Figure 1.10). Building Information Modeling (BIModeling) refers to a multidisciplinary and collaborative process to manage a built asset over various phases of its lifecycle (Eastman et al. 2011). The BIM process leverages and leads to the production of a Building Information Model (BIModel), which is a structured 3D spatial dataset(s) describing all of the physical components of the built asset over its lifespan (Eastman 1999).

BIM products provide an efficient spatial representation, as well as communication of architectural and structural elements inside complex

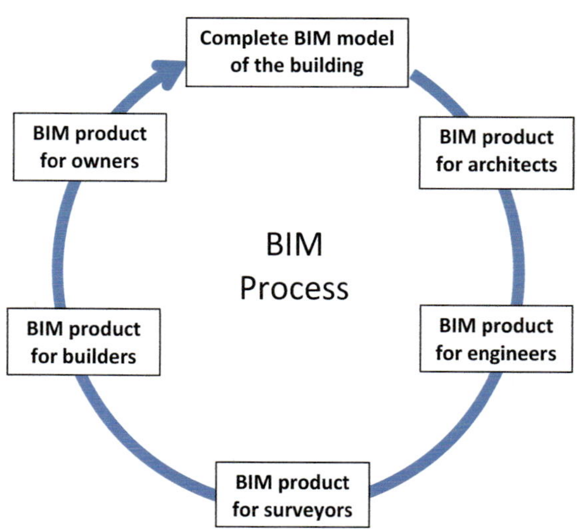

FIGURE 1.10
BIM: Process and product perspectives.

built assets. There are multiple players involved in the development of any built asset; these include architects, engineers, builders, surveyors, and owners, and each player can create and use their unique BIM product. During the course of the BIM process, these distinct products can be combined in a common data environment which provides a complete model of the built asset. BIM products not only include tangible spatial objects, such as windows, walls, doors, and slabs, but can also contain intangible volumetric spaces. In addition, some spatial relationships between tangible and intangible spatial objects are defined within BIM models (Borrmann and Rank 2009).

The Industry Foundation Classes (IFC) standard is the open data model for exchanging BIM models across multiple BIM platforms, and has been developed for promoting interoperability over the development process of built assets (ISO16739 2013). IFC is an object-oriented data model that includes a complete set of entities describing geometric and semantic aspects of a built asset throughout its lifecycle. This standard typically uses a hierarchical spatial structure in order to store building information. As a result, the BIM models in an IFC format can be decomposed into manageable subsets such as buildings, sites, stories, spaces, and building elements inside stories (Liebich 2009).

1.7 Current Benefits and Challenges of BIM in the AEC Industry

In the AEC industry, BIM has had a major impact on resolving issues associated with 2D drawings extending to 3D CAD models (Arayici et al. 2011; Mihindu and Arayici 2008). A brief review of these benefits is presented in the following (Azhar 2011; Ballesty et al. 2007):

- Accurate representation of complex buildings: One of the key advantages of BIM is that it can provide accurate representation of the geometry of various parts of complex buildings within an interactive and integrated data environment.
- Accelerating development process of buildings: Since a BIM environment facilitates sharing and reusing building information, building development actors can easily collaborate with each other, which would in turn increase the speed of design and construction.
- Facilitating design tasks: Proposed designs for buildings can be evaluated faster in terms of performing and benchmarking simulations and analyses. For example, a BIM environment would improve "daylighting" analysis of a building development. This would also result in enhanced and innovative design proposals.

- Costs and time savings: According to an investigation conducted by the Centre for Integrated Facilities Engineering at Stanford University (Gao and Fischer 2008), the unbudgeted change in construction projects can be eliminated by up to 40 percent through the adoption of a BIM-based paradigm. In addition, the accuracy of estimating costs is around 3 percent, which is more accurate in comparison with traditional methods of estimation. The time spent to estimate the costs of project can be reduced by up to 80 percent. Moreover, the time to complete each project was reduced by up to 7 percent.

More recently, Barlish and Sullivan (2012) suggested a valid framework or baseline comprising two main categories of metrics for measuring the potential benefits of BIM in AEC projects. The first category is the return metrics, namely change orders, RFIs (Requests for Information), and schedule. The second category is investment metrics, including design costs and contractor costs. These metrics were measured using three real-world case studies. Their main finding was that benefits and cost savings of the BIM paradigm outweighs its upfront cost and risk of rework over a long period of time. It was also found that the BIM paradigm relies on various factors to be successful in AEC projects (Barlish and Sullivan 2012).

Some important factors include project size, proficiency of staff members in BIM, communication of the project team, as well as other organizational external factors. This indicates that success of a BIM-based paradigm differs from one project to another. Finally, it was argued that the full potential of a BIM-based paradigm is yet to be realized, which stems from the fact that there is no sufficient business case of this paradigm for AEC actors (Barlish and Sullivan 2012).

Bryde, Broquetas, and Volm (2013) indicated that the most frequently stated benefit of a BIM-based paradigm is cost and time reduction, as well as enhancing control of tasks and activities over the lifecycle of AEC projects. Other benefits reported were communication improvement and coordination enhancement (Bryde, Broquetas, and Volm 2013). Communication improvement indicates that finding specific information is much easier in an integrated data environment of BIM compared to the fragmented environment of traditional 2D drawings. Additionally, the changes in BIM models are better communicated to the owners. Coordination enhancement implies that using BIM would improve workflows and would facilitate integrated design strategies and enable Integrated Project Delivery (IPD).

While BIM contributes to the many benefits for productivity within the AEC industry, there are still challenges associated with the adoption of BIM that need to be still addressed. These challenges can be classified into two main categories, namely, institutional and technical (Azhar 2011). One institutional challenge is to resolve issues associated with ownership of data in a BIM environment, which requires the protection of BIM data via copyright laws or similar legal channels (Bryde, Broquetas, and Volm 2013).

For instance, if the design model is bought by the owner, he/she may think that complete ownership of this model is entitled to him/her. However, there might be some proprietary information, within the model, which is entitled to the design team. This proprietary information must be protected when owners use the model in the project.

There is no straightforward solution for resolving issues associated with ownership of BIM data, and it may differ from one project to another in accordance with the requirements of project participants (Azhar 2011). One possible solution to avoid copyright issues could be clearly specifying various rights associated with the ownership of BIM data inside the contract documents (Rosenberg 2007). Another legal issue is to determine a responsible party for controlling the accuracy of the data (Thompson and Miner 2006). Updating BIM data and monitoring its accuracy is inevitably associated with taking risks. This also results in spending more time on preparing and versioning BIM data, which can be considered as a new cost over the development process of the project. Although this cost will be compensated by efficiency gains, it is still a cost that will be incurred within the project. Before full implementation of a BIM-based paradigm, these institutional issues should be appropriately identified.

Bernstein and Pittman (2004) specify three broad categories for technical challenges of a BIM-based paradigm, namely transactional business process evolution, computability of digital design information, and meaningful data interoperability. Transactional business process evolution indicates that there is a need to define appropriate business process models to facilitate data integration within a BIM environment. Computability of digital design information indicates the capability of a BIM environment to automatically compute various quantities (such as measuring volumes and areas of spaces or rooms defined within the model) from digital data. Therefore, there is a need to move from pictorial BIM products to computable ones.

The most critical technical challenge is the interoperability between varying BIM tools, as each tool has its own unique data format for manipulating BIM data (Isikdag et al. 2007). This challenge can be addressed by developing an open data model which facilitates data transfer issues between different BIM tools (Eastman et al. 2011). The international BuildingSMART organization has since developed the IFC standard to facilitate interoperability between BIM tools (Liebich 2013).

1.8 Potential of BIM for Urban Land Administration

BIM models include highly detailed physical information about built assets. A BIM environment has the potential to considerably advance land administration systems in urban areas, since it captures 3D digital data associated with the planned and real built environment. As a rich 3D digital environment of buildings, BIM models can provide a significant data input for urban land administration practices. However, these models currently do not directly implicate legal information associated with the ownership of built assets (Atazadeh et al. 2017). There is no adequate interplay between the BIM environment and land administration systems in current building subdivision practices of most jurisdictions around the world. This may stem from the fact that a BIM-based paradigm was initially created for designing architectural and structural aspects of buildings.

Therefore, the potential value of BIM and in particular the IFC data model has recently become a focus of various investigations in the land administration domain (e.g. Clemen and Gründig 2006; El-Mekawy and Östman 2012; Isikdag et al. 2015). Although these recent studies showcase various levels of using BIM for assisting land administration functions, there are no studies that have explored how the IFC standard can be extended to accommodate legal information, which plays an underpinning role in adopting BIM for urban land administration purposes.

Encapsulation of legal data elements within the BIM environment would provide a feasible solution for addressing current spatial and communication problems in urban land administration (Atazadeh et al. 2016; Atazadeh, Rajabifard, and Kalantari 2017). BIM models can potentially add value to the current processes of recording, registering, and visualizing legal information within complex multi-story developments. With this in mind, the aim of this book is to describe how the BIM environment and its open data model (IFC) can be extended to incorporate legal information, as a primary step towards leveraging BIM for the 3D digital management of land and property information inside and around multi-story developments (see Figure 1.11).

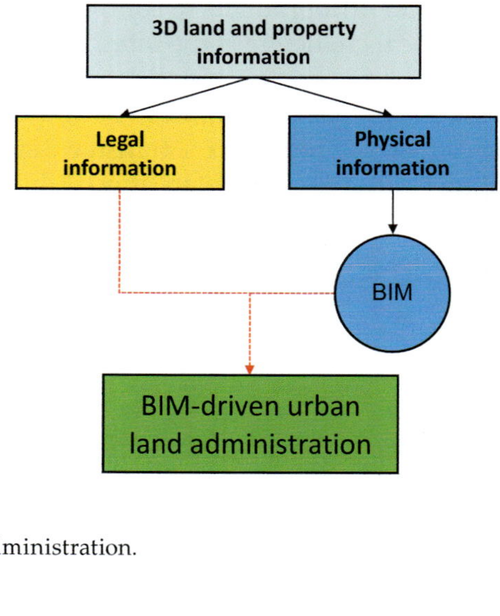

FIGURE 1.11
BIM for urban land administration.

References

Acharya, B. R. 2011. "Prospects of 3D Cadastre in Nepal." In: *2nd International Workshop on 3D Cadastres*, 16–18 November 2011, Delft, the Netherlands.

Arayici, Y., P. Coates, L. Koskela, M. Kagioglou, C. Usher, and K. O'Reilly. 2011. "Technology Adoption in the BIM Implementation for Lean Architectural Practice." *Automation in Construction* 20 (2):189–195.

Atazadeh, Behnam, Mohsen Kalantari, Abbas Rajabifard, Tom Champion, and Serene Ho. 2016. "BIM for 3D Digital Management of Stratified Ownership Rights in Buildings." In: *FIG Working Week 2016 Recovery from Disaster*, 2–6 May, Christchurch, New Zealand.

Atazadeh, Behnam, Abbas Rajabifard, and Mohsen Kalantari. 2017. "Assessing Performance of Three BIM-Based Views of Buildings for Communication and Management of Vertically Stratified Legal Interests." *ISPRS International Journal of Geo-Information* 6 (7):198.

Atazadeh, Behnam, Mohsen Kalantari, Abbas Rajabifard, Serene Ho, and Tom Champion. 2017. "Extending a BIM-Based Data Model to Support 3D Digital Management of Complex Ownership Spaces." *International Journal of Geographical Information Science* 31 (3):499–522.

Azhar, Salman. 2011. "Building Information Modeling (BIM): Trends, Benefits, Risks, and Challenges for the AEC Industry." *Leadership and Management in Engineering* 11 (3):241–252.

Ballesty, S., J. Mitchell, R. Drogemuller, H. Schevers, C. Linning, G. Singh, and D. Marchant. 2007. "Adopting BIM for Facilities Management: Solutions for Managing the Sydney Opera House." 1–32. Cooperative Research Centre (CRC) for Construction Innovation: Brisbane, Australia.

Barlish, Kristen and Kenneth Sullivan. 2012. "How to Measure the Benefits of BIM—A Case Study Approach." *Automation in Construction* 24 (July):149–159.

Bernstein, Phillip G. and Jon H. Pittman. 2004. *Barriers to the Adoption of Building Information Modeling in the Building Industry*. Autodesk Building Solutions.

Biljecki, Filip, Hugo Ledoux, and Jantien Stoter. 2016. "An Improved LOD Specification for 3D Building Models." *Computers, Environment and Urban Systems* 59:25–37.

Borrmann, André and Ernst Rank. 2009. "Specification and Implementation of Directional Operators in a 3D Spatial Query Language for Building Information Models." *Advanced Engineering Informatics* 23 (1):32–44.

Bryde, David, Martí Broquetas, and Jürgen Marc Volm. 2013. "The Project Benefits of Building Information Modelling (BIM)." *International Journal of Project Management* 31 (7):971–980.

Clemen, Christian and L. Gründig. 2006. "The Industry Foundation Classes (IFC)–Ready for Indoor Cadastre?" In: *Proceedings of XXIII International FIG Congress*. Vol. 18, 8–13 October 2006, Munich.

Döner, Fatih, Rod Thompson, Jantien Stoter, Christiaan Lemmen, Hendrik Ploeger, Peter van Oosterom, and Sisi Zlatanova. 2010. "4D Cadastres: First Analysis of Legal, Organizational, and Technical Impact—With a Case Study on Utility Networks." *Land Use Policy* 27 (4):1068–1081.

Eastman, Charles M. 1999. *Building Product Models: Computer Environments, Supporting Design and Construction*. Boca Raton, FL: CRC Press.

Eastman, Charles M., Paul Teicholz, Rafael Sacks, and Kathleen Liston. 2011. *BIM Handbook: A Guide to Building Information Modeling for Owners, Managers, Designers, Engineers and Contractors*. John Wiley & Sons.

El-Mekawy, Mohamed and Anders Östman. 2012. "Feasibility of Building Information Models for 3D Cadastre in Unified City Models." *International Journal of E-Planning Research* 1 (4):35–58.

Gao, Ju and Martin Fischer. 2008. *Framework and Case Studies Comparing Implementations and Impacts of 3D/4D Modeling Across Projects*. Stanford University, CA: Center for Integrated Facility Engineering.

García, José Miguel Olivares, Luis Ignacio Virgós Soriano, and Amalia Velasco Martín-Varés. 2011. "3D Modeling and Representation of the Spanish Cadastral Cartography." In: *2nd International Workshop on 3D Cadastres*, 16–18 November 2011, Delft, the Netherlands.

Groger, G., Thomas Kolbe, Claus Nagel, and K. H. Hafele. 2012. *OGC City Geography Markup Language (CityGML) En-Coding Standard*. Wayland, MA: Open Geospatial Consortium.

ICSM. 2010. "EPlan Model." The Intergovernmental Committee on Surveying and Mapping, Canberra, Australia. Available at: www.icsm.gov.au/sites/default/files/ICSM-ePlan-Model-v1.0 %281%29.pdf.

Isikdag, Umit. (Ed.) 2015. "Building Information Models: An Introduction." *Enhanced Building Information Models – Using IoT Services and Integration Patterns. Springer Briefs in Computer Science*, 1–12. Springer International Publishing. doi:10.1007/978-3-319-21825-0_1.

Isikdag, Umit, Ghassan Aouad, Jason Underwood, and Song Wu. 2007. "Building Information Models: A Review on Storage and Exchange Mechanisms." In: *24th CIB W78 Conference*, 26–29 June 2007. Maribor, Slovenia.

Isikdag, Umit, Mike Horhammer, Sisi Zlatanova, R Kathmann, and Peter van Oosterom. 2015. "Utilizing 3D Building and 3D Cadastre Geometries for Better Valuation of Existing Real Estate." In: *FIG Working Week 2015: From the Wisdom of the Ages to the Challenges of the Modern World*, 1–18, 17–21 May 2015, Sofia, Bulgaria.

ISO16739. 2013. "Industry Foundation Classes (IFC) for Data Sharing in the Construction and Facility Management Industries." International Organization for Standardization.

ISO19152.2012. "Geographic Information – Land Administration Domain Model (LADM)." International Organization for Standardization.

Jazayeri, Ida, Abbas Rajabifard, and Mohsen Kalantari. 2014. "A Geometric and Semantic Evaluation of 3D Data Sourcing Methods for Land and Property Information." *Land Use Policy* 36:219–230.

Kolbe, Thomas. 2009. "Representing and Exchanging 3D City Models with CityGML." In: Lee, Jiyeong Lee, and Zlatanova, Siyka, *3D Geo-Information Sciences*, 15–31. Springer.

Land Title Act. 1994. *Digital Amigo*. Available at: http://www.digitalamigo.com.au/WEB%20SITE/PDFs/Volumetric%20Example%20Sh%209.pdf.

Libbis, Simon. 2015. *Subdivisions Victoria: The Ultimate Guide*. Melbourne, Australia: Hybrid Publishers.

Liebich, T. 2009. *IFC 2x Edition 3. Model Implementation Guide. Version 2.0*. AEC3 Ltd.

Liebich, T. 2013. *IFC4–the New BuildingSMART Standard*. buildingSMART.

Mihindu, S. and Y. Arayici. 2008. "Digital Construction through BIM Systems Will Drive the Re-Engineering of Construction Business Practices." *Information Visualisation, 2008 International Conference*. IEEE: London, UK.

NBIMS. 2012. *National BIM Standard – United StatesTM. Version 2*. Available at: http://www.nationalbimstandard.org/nbims-us-v2/pdf/NBIMS-US2_aB.pdf.

Rajabifard, Abbas, Behnam Atazadeh, and Mohsen Kalantari. 2018. "A Critical Evaluation of 3D Spatial Information Models for Managing Legal Arrangements of Multi Owned Developments in Victoria, Australia." *International Journal of Geographical Information Science* 32 (10):2098–2122.

Rajabifard, Abbas, Ian Williamson, Brian Marwick, Mohsen Kalantari, Serene Ho, Davood Shojaei, Behnam Atazadeh, Sam Amirebrahimi, and Alireza Jamshidi. 2014. "3D-Cadastre, a Multifaceted Challenge." In: *The XXV FIG International Congress 2014 Engaging the Challenges, Enhancing the Relevance*, 16–21 June 2014, Kuala Lumpur, Malaysia.

Rosenberg, Thomas L. 2007. "Building Information Modeling." *Partnering for Success*. Available at: https://www.academia.edu/1283324/Building_Information_Modeling.

Shojaei, Davood. 2014. "3D Cadastral Visualisation: Understanding Users' Requirements." Ph.D. thesis, University of Melbourne, Australia. Available at: http://hdl.handle.net/11343/45129.

Shojaei, Davood, Abbas Rajabifard, Mohsen Kalantari, Ian D. Bishop, and Ali Aien. 2014. "Design and Development of a Web-Based 3D Cadastral Visualisation Prototype." *International Journal of Digital Earth*, 1–20.

Stoter, Jantien, Peter van Oosterom, Hendrik Ploeger, and Henri Aalders. 2004. "Conceptual 3D Cadastral Model Applied in Several Countries." In: *FIG Working Week*. Vol. 2004, Athens, Greece.

Thompson, D. B. and Ryan G. Miner. 2006. "Building Information Modeling-BIM: Contractual Risks Are Changing with Technology." Available at: www.aepronet.org/ge/no35.html.

Vandysheva, Natalia, Sergey Sapelnikov, P. J. M. Van Oosterom, M. E. De Vries, Boudewijn Spiering, Rik Wouters, Andreas Hoogeveen, and Veliko Penkov. 2012. "The 3D Cadastre Prototype and Pilot in the Russian Federation." In: *Proceedings FIG Working Week 2012, Territory, Environment and Cultural Heritage*, 6–10 May 2012, Rome, Italy. International Federation of Surveyors (FIG).

Victoria State Government. 2019. "Example – Multi-Story Building Subdivision." Available at: https://www.propertyandlandtitles.vic.gov.au/__data/assets/word_doc/0028/53875/5.-Multi-storey.docx.

Williamson, Ian P., Stig Enemark, Jude Wallace, and Abbas Rajabifard. 2010. *Land Administration for Sustainable Development*. Redlands, CA: ESRI Press Academic.

2

Urban Land Administration—A Digital Paradigm

2.1 Introduction

Urban areas are characterized by the prolific existence of multi-story building structures, complex shopping centers, complex aboveground and underground transport networks, parking spaces, and utility networks. These distinctive features of cities have resulted in a wide variety of rights, restrictions, and responsibilities (RRR) associated with the legal ownership and use of properties in urban areas. The legal boundaries of urban properties typically intersect with complex structural and architectural components; this makes it very challenging to identify the precise spatial location of legal boundaries inside complex urban developments. Urban land administration refers to the information and processes required to support subdivision, registration, and the ongoing management of private, communal, and public legal interests vertically located in complex urban areas (Ho 2014). Figure 2.1 represents a good example of complex urban developments which comprise a wide range of private, communal, and public legal interests.

The traditional paradigm for urban land administration, as part of wider land administration systems, was predicated on devising 2D-based notions for recording, managing, and representing spatial dimensions of 3D RRR spaces in vertical urban developments. As presented in Section 1.2, this 2D-based paradigm faces a wide variety of challenges associated with the communication and management of spatial complexities in vertically located properties. In order to address these challenges, 3D digital data environments are increasingly adopted as a new paradigm in urban land administration. The practice of urban land administration is also predicated on the legislative framework of a jurisdiction. The elements of urban land administration are very much dependent on the legislation, regulations, and guidelines that are adopted by a specific jurisdiction to define legal boundaries and spatial arrangements of 3D RRR spaces. Therefore, in this chapter, a brief review of urban land administration paradigms in selected jurisdictions around the world will be presented. Following a brief review, the elements of urban land

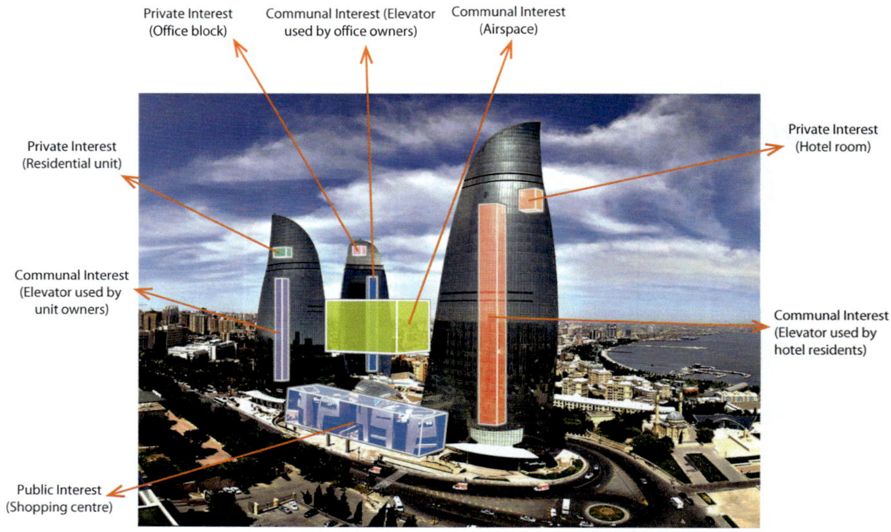

FIGURE 2.1
An example of a complex urban development comprising various private, communal, and public properties. (Flame towers in Baku, Azerbaijan.)

administration in the state of Victoria in Australia will be described in more detail. Finally, 3D spatial data models adopted in digital urban land administration will be introduced. The foundation of a digital paradigm for urban land administration can be achieved by adopting a 3D spatial data model, since they provide a coherent structure for storing 3D digital data in land and property management systems.

2.2 Urban Land Administration—An International Context

Each country or state has adopted a unique approach to recording and managing RRR spaces associated with private, communal, and public properties vertically located in urban environments. The legal definition of arrangements and boundaries of 3D RRR spaces plays an underpinning role in urban land administration practices devised by a jurisdiction. There have been some considerable research efforts focusing on the digitization of 3D RRR spaces. Researchers from around the world have studied concepts and methods of urban land administration with respect to their juridical framework. An overview of urban land administration in five jurisdictions selected from various parts of the world will be provided. These jurisdictions include Canada (Québec), Malaysia, The Netherlands, Sweden, and Australia (Victoria). The reason for selecting these jurisdictions is that they have initiated and conducted significant projects to realize the digital paradigm for urban land administration in line with their own jurisdictional settings. Hence, a brief description of 3D-enabled urban land administration activities in these jurisdictions will be also provided in the following subsections.

2.2.1 Canada

In the Québec province, two types of legal interests underpin the basis of urban land administration practices. The first type of legal interest is called *immatriculated* legal objects, which are typically private lots and common properties associated with land parcels, condominium apartments, and any 3D physical object (Pouliot et al. 2011). These legal interests are registered in the land registry system. In simple developments, the spatial extent of these legal interests is represented in a 2D cadastral map, in which the lot number and its dimensions are also annotated alongside its boundary lines. In complex and vertical urban developments, a particular scheme called "cadastre vertical" is used to define vertical limits of private and communal properties. In this case, the land parcel on the 2D cadastral map references a complementary or supplementary plan (PC). Complementary plans are typically stored as PDF files, and include an overall view of the building, 2D floorplans as well as cross-section diagrams. Private lots are only annotated with measurements alongside their boundary lines in three dimensions. Communal parts of a building are only annotated with a lot number in complementary plans.

The second type of legal interest is defined in other registries which are complementary to the land registry system. Typical examples of these legal interests are state resources and private utility systems (Pouliot, Cuissart and Bordin 2015). These types of legal interests are not spatially referenced making it very difficult for the responsible authorities to determine their spatial location.

Pouliot et al. (2011) conducted the first experiments to support the move towards digital urban land administration in the province of Quebec. They proposed a workflow for creating 3D digital models of vertically stratified properties from PC files. This workflow comprises three main stages: vectorization, 3D modeling, and data exchange. In the vectorization stage, the images of each lot are transformed to geospatially referenced vector formats. The converted vector files include lots as closed polylines ready for 3D digital modeling. In the 3D modeling phase, the polylines for each lot are converted to 3D spatial objects using simple extrusion and Constructive Solid Geometry (CSG) approaches (see Figure 2.2). Finally, in the data exchange phase, some tests were conducted for importing and exporting 3D digital models between various 3D tools such as ArcScene, FME Data Inspector, 3D PDF, and Google Earth. The outcome of the proposed workflow would provide the users with better access to 3D digital data for the purpose of urban land administration.

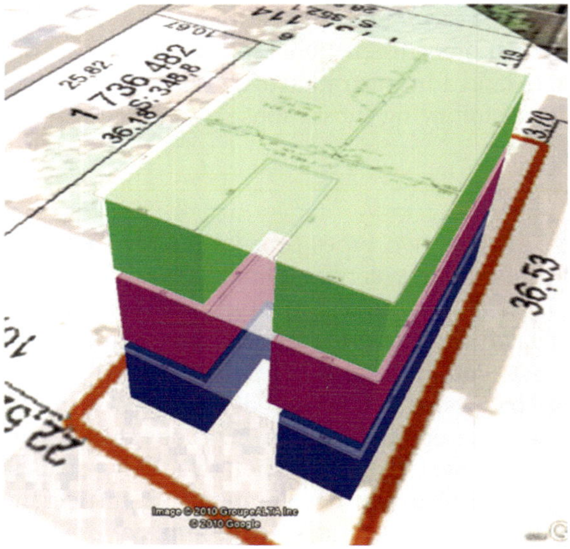

FIGURE 2.2
An implemented 3D digital model of apartment lots. (Adapted from Pouliot et al. 2011.)

2.2.2 Malaysia

Malaysian urban land administration practices use two types of plans for managing multilayered properties: strata plans and stratum plans. Strata plans provide the spatial dimensions of assets and properties located above the ground (see Figure 2.3). These plans provide legal spaces and legal boundaries for each level of a multi-story development. In strata titles, various legal interests are defined, including the land parcel, the parcel unit, common property, and the accessory unit. The land parcel in this case refers to a strata unit which is defined inside a master lot with a building with no more than four stories. A parcel unit refers to an individual unit within an apartment or condominium. Parcel units are represented in white with red boundaries. An accessory unit is a unit which is shown in a strata plan and will be used in conjunction with a parcel unit. These are represented in white and their boundaries are delineated in blue.

A common property refers to those parts of a master lot which are not included in any individually owned unit, including accessory units. Limited common properties must be exclusively used by a specific group of owners of strata lots. Common properties are represented in yellow in strata plans. Legal boundaries define the spatial extent of each legal interest. These boundaries in strata plans are boundaries that are specified based on surveying measurements such as distance, angle, and azimuth. Within apartments, legal boundaries are defined as lines passing through the median of physical structures, such as walls and ceilings. In strata plans, legal boundaries for

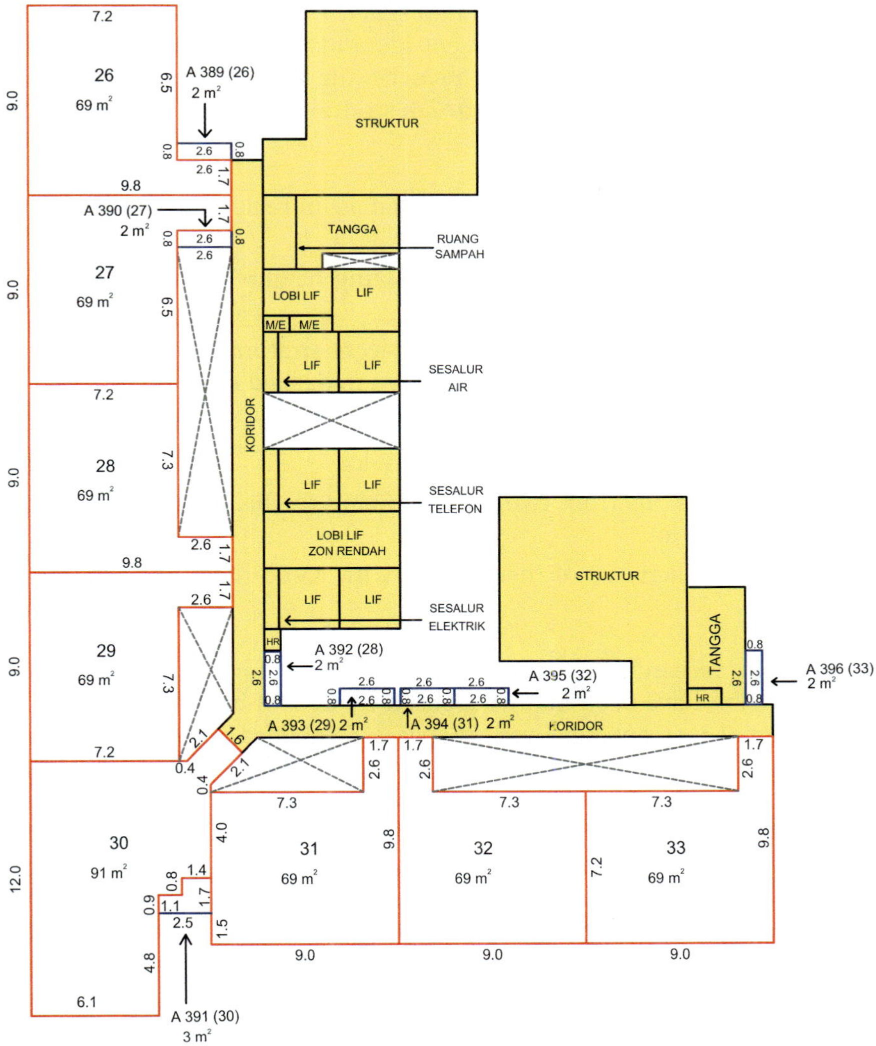

FIGURE 2.3
An example diagram from a strata plan in Malaysian jurisdiction.

parcel units, accessory units, and common properties are represented in red, blue, and black, respectively.

Stratum plans provide a volumetric representation of underground properties. For this purpose, underground stratum alienation involves defining a legal volumetric ownership of a property at a specified depth under the surface of the earth. Stratum plans are prepared based on as-built surveys comprising three types of diagrams, namely, a location diagram, a usage diagram, and a stratum diagram. Location diagrams depict the location in which the stratum structure is constructed according to a suitable scale based on the cadastral map or standard sheet. This diagram should include information about title, scale, adjacent lot, and direction. Usage diagrams provide information about the heights, schedule of stratum utilizations, and floorplans. As represented in Figure 2.4, stratum diagrams contain the following items:

- Bearing and distance of the lot boundary in relation to the stratum structure.
- Position of two (2) benchmarks showing heights with reference to the vertical reference datum.
- Position of adjacent lots and the adjacent stratum lots (if any). The adjacent stratum lots should be shown in dashed lines.
- Name and offsets to the road (if any).
- Cross-sections to indicate the stratum lot.
- Vertical sections from two sides showing the heights/elevations of the stratum lot.
- Isometric diagrams of the stratum with coordinates and volumes shown.

In Malaysia, research and development activities for digital urban land administration have been mainly focused on LADM-based approaches for 3D digital management of ownership spaces. Rajabifard et al. (2018) proposed potential solutions for implementing a 3D national digital cadastral database (3D-NDCDB), which included the processes for upgrading the existing dataset and data collection methods to support the 3D digital data and the creation of a 3D spatial database based on the elicited user requirements.

The implementation of 3D-NDCDB was considered as an initial step to developing a 3D urban land administration system that includes the upgrade of the hardware, software, and application modules to support 3D digital management of land and property. The major outcome was an open-source 3D spatial database, which is LADM compliant, to address issues with regards to the existing eKadastre project in Malaysia.

2.2.3 The Netherlands

Land administration in the Netherlands adopts a deed-based approach for registering and managing properties (Stoter 2004). This approach was introduced in the early years of the 19th century. Register of deeds, cadastral registration, and cadastral base map are maintained by the Netherlands Cadastre and Land Registry Organization. The legal system in the Netherlands to define the legal ownership of a piece of land and its underground and aboveground properties is based on Roman law (Kitsakis et al. 2018). In Dutch jurisdiction, the legal ownership of a land parcel not only comprises

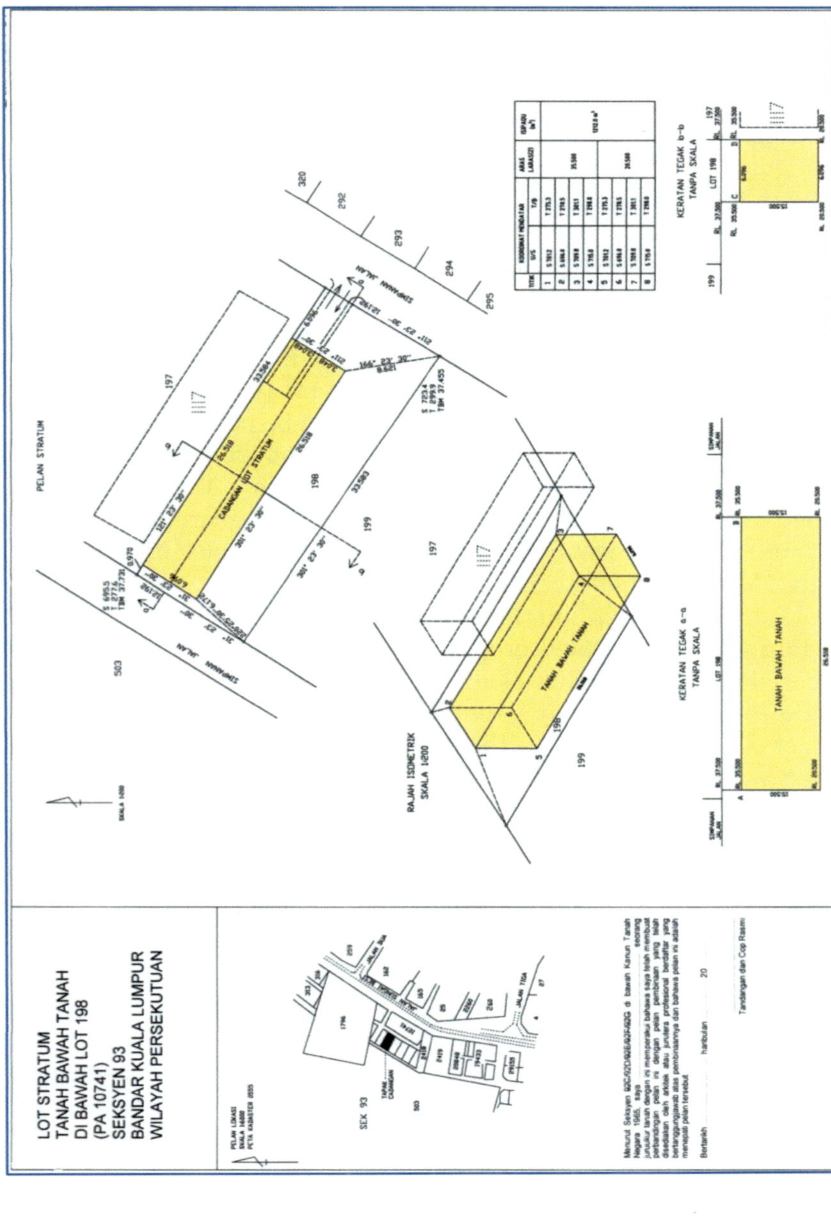

FIGURE 2.4
An example of a stratum diagram in Malaysian jurisdiction.

the land parcel itself but also any property that is located in below or above the land parcel. These properties include buildings, infrastructure facilities, subsurface assets, and the airspace above the land parcel.

Unlike the common law system, Dutch law does not allow the subdivision of the legal ownership of a land parcel into 3D volumetric legal spaces. However, in the context of urban land administration, the legal framework in the Netherlands allows the land owners to define vertically stratified property rights by establishing condominium rights (apartment rights) and restricted rights affecting the land parcel (Stoter, Ploeger and van Oosterom 2013). In Dutch jurisdiction, there is also another rule known as "horizontal accession to real estate." This rule is applied to those situations where a building is developed on two land parcels with separate ownership rights. In these cases, the legal ownership of the building is not communally owned by the owners of land parcels; it is entitled to the owner whose land parcel includes the major part of the building.

In the Netherlands, using condominium rights is a typical approach to allow the separate use of 3D legal spaces within a multi-story development (Stoter, Ploeger, and van Oosterom 2013). It should be noted that the entire development, which comprises the land parcel and buildings, is jointly owned by all owners. Each apartment right, which is a 3D legal space, is considered as a share in this communal ownership and gives the right to its owner to use the apartment exclusively. All the owners must be also members of the Homeowners Association for the development.

In addition to condominium rights, restricted rights are sometimes used for creating vertically located properties. These restricted rights include superficies, long lease, and easement. A right of superficies provides the legal right to establish a built asset below or above a land parcel owned by another owner. Interestingly, the interest holder of this restricted right is entitled with the legal ownership of the built asset located on a land parcel. Hence, the right of superficies is another approach to separate the legal ownership of an aboveground or underground property from its land parcel. Its spatial extent is therefore clearly defined by 3D volumetric spaces. Examples of built assets owned by right of superficies include underground carparks, road and metro tunnels, utility networks, and wind turbines.

The right of long lease provides the right to use, rather than own, a property. The concept of this right is not the same as the lease right in the common law system. The right of long lease is a restricted right in rem; however, the lessee is economically in a position which is almost the same as the legal owner of the property. This right is typically applied to both the land and any property located below and above the land. However, in some case, a volume of space is also specified as a right of long lease. For instance, a volume of space above a metro tunnel can be leased for the purpose of establishing buildings. An easement is a restriction on the serving parcel for the benefit of the owner of another parcel in the neighborhood of the serving parcel. Typical examples of an easement are a right of way and a right to use a pipeline crossing over the serving parcel.

The 3D spatial extent of condominium rights is defined based on a specific requirement determined in the Dutch law. According the requirements, a separate plan for each floor of an apartment should be prepared and registered in the land registry. The boundaries of apartment units are clearly delineated in the floorplans. There are no particular requirements in the Netherlands for demarcating the 3D spatial extent of restricted rights. The general principle is that the spatial extent of these rights should be "identifiable."

The current practice of urban land administration in the Netherlands is challenged by it requiring substantial mental exercise to understand ownership rights in spatially complex situations. Therefore, Stoter, Ploeger, and van Oosterom (2013) proposed a two-phase solution for 3D digital enabled urban land administration. The first phase is in line with current cadastral and legal frameworks and can be implemented in the short term. The initial result of the first phase is 3D digital drawings representing volumetric legal spaces. In this phase, BIM models were also used for 3D digital registration of properties; this will be explained in Section 3.5.5.

The major drawback of the first phase is that it provides a purely visualization solution. It does not provide direct access to 3D digital data, which comprises the geometry and semantics of 3D legal spaces, in order to check the spatial validity of 3D legal spaces. Therefore, in the second phase, the proposed approach was used to enable 3D-digital urban land administration for the entire Dutch jurisdiction in the long term. This phase is still under investigation and it aims to provide a fully 3D digital environment for urban land administration. It would lead to fundamental changes in the principles currently adopted for land and property registration. A practical solution for this phase should be developed based on a range of technical, juridical, and organizational factors.

2.2.4 Sweden

In Sweden, urban land administration practices had been traditionally predicated on 2D property units. In 2004, the concept of 3D property was introduced in the Swedish Land Code (Larsson, Paasch and Paulsson 2018). In Swedish jurisdiction, a 3D property is defined as a property unit which is delimited in both horizontal and vertical dimensions (Paulsson 2007). 3D property is used for recording and managing legal rights associated with different urban land uses, such as multi-story developments comprising residential units, office blocks, and commercial premises. In addition, it can be also adopted for defining legal ownership of infrastructure assets, such as tunnels and underground facilities.

The Swedish legislation does not specify particular requirements for delineating legal boundaries with respect to physical structures. The suitable location of legal boundaries is determined based on the existing conditions in each development. The concept of 3D property was later extended with the introduction of condominium (apartment) ownership in 2009. Condominium ownership is defined as a specific type of 3D property unit, which is used for representing ownership rights in residential developments.

The 3D property unit must define the legal ownership of the whole, or part, of a built asset. The spatial extent of a 3D property unit is not necessarily defined inside the legal boundaries of 2D property units. 3D property units are registered in the same as way as the 2D property units. However, additional 3D spatial information is provided in the land registry and cadastral index map. This information is typically provided as textual notations, which include horizontal (x, y) coordinates and height information. Visual representation of 3D property units in the Swedish cadastral index map relies on basic 2D drawings (see Figure 2.5). The legal boundaries are represented by dashed lines (Larsson, Paasch and Paulsson 2018). Special texture and cartographic annotations are used to highlight the area covered by a 3D property.

The current approach of urban land administration in Sweden does not provide an effective mechanism to reuse 3D spatial information associated with 3D properties. During the land administration process, there is access to 3D construction models with good quality; however, these models are used for creating 2D drawings and textual 3D information. The current procedures do not provide an integrated 3D data environment for managing 3D properties. Therefore, there have been research projects conducted in Sweden to visualize and standardize 3D digital information for urban land administration (Andrée et al. 2018). These investigations are part of the Swedish "Smart Built Environment" program which aims to adopt the BIM environment for 3D property formation processes. From a visualization perspective, some

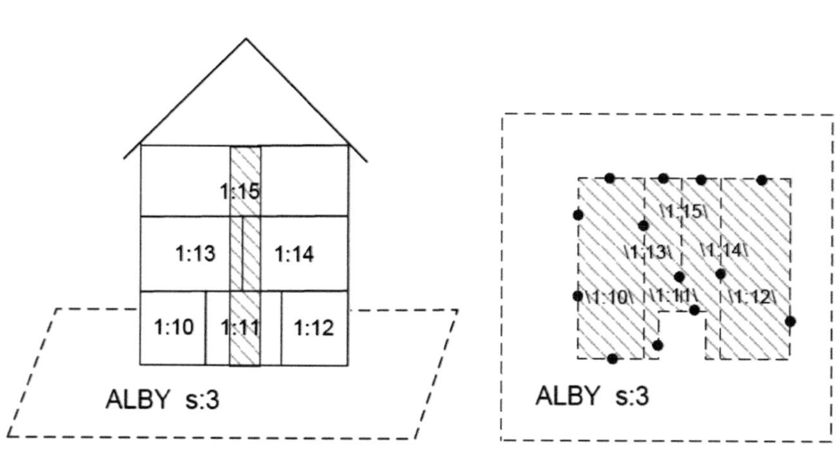

FIGURE 2.5
A cross-section diagram of 3D property units in a multi-story building (left) and its visualization on the digital cadastral index map (right). (Adapted from Larsson, Paasch, and Paulsson 2018.)

pilot studies were conducted to represent 3D property rights and boundaries using BIM models. However, these studies are still in the early stages and require further practical tests. For standardization of the 3D digital property information, research projects were conducted on extending the Unified Building Models (UBM) with property boundaries (El-Mekawy and Östman 2012). This will be explained further in Section 3.5.3.

2.2.5 Australia

Australia comprises eight jurisdictions including six states: New South Wales, Queensland, South Australia, Tasmania, Victoria, Western Australia, and two mainland self-governing territories: the Australian Capital Territory and the Northern Territory. Currently, urban land administration practices in Australian jurisdictions mainly rely on 2D analogue subdivision plans to define the legal boundaries of properties. However, each jurisdiction has its own unique approach for subdividing the legal ownership of urban properties.

For example, in Queensland, volumetric lots delineated on 2D plans are used to represent the spatial extent of legal arrangements of multi-owned developments, while floorplans and cross-section diagrams are used in New South Wales and Victoria to define the legal ownership of properties within multi-owned developments. Among the Australian jurisdictions, Victoria is a good choice for understanding current urban land administration practices due to three main reasons. First, Victoria is the most populous state, with many urban developments. In addition, the legal boundary types defined in the Victorian jurisdiction are comprehensive, and all the other jurisdictions include the same number, or a subset, of these legal boundary types. Finally, Victoria is the leading state in terms of investigating digital solutions for urban land administration in Australia (Olfat et al. 2018).

2.3 Urban Land Administration in Victoria, Australia

In this section, urban land administration in Victoria, Australia, will be investigated thoroughly to demonstrate the procedures and data elements of urban land administration in the context of one specific jurisdiction. This will provide the basis for the next chapters of the book, in which we will explain how the digital environment of BIM can be utilized in the context of urban land administration. Although the BIM-driven urban land administration described in this book is mainly focused on the Victorian jurisdiction, this approach can be similarly adopted for any jurisdiction around the world.

Urban land administration is a complex process that involves various entities such as actors, tasks, and documents. Business Process Modeling Notation (BPMN) can be used to facilitate an understanding of this complex process (White and Bock 2011). Diagrammatic maps created using BPMN notation represent the whole lifecycle of any business process. Readers are referred to Appendix B which provides an informative description of BPMN notations used in this book. The lifespan of a typical process for urban land administration in Victoria, Australia mainly includes four phases, namely planning, certification, compliance, and registration. Each phase will be described in detail in the following subsections.

2.3.1 Planning Phase

The developer or owner starts the land development process by identifying a suitable piece of land. Once the suitable piece of land has been identified, the land surveyor conducts a site survey to establish the legal boundaries of the land. Meanwhile, an architect constructs the design model of the building and its architectural elements. Once the developer agrees on the design model of the building, the land surveyor uses the architectural structure of the building as the basis for preparing the subdivision plan of the building. Land surveyors should prepare subdivision plans in alignment with the Subdivision Act (State Government of Victoria 1988) and Regulations (State Government of Victoria 2011b). In preparing subdivision plans, land surveyors define various types of legal boundaries inside buildings. Subsequently, they propose a range of legal arrangements such as individually owned lots, common properties, and easements. Finally, if there is a need to create an Owners Corporation, they propose an entitlement and liability value for each individually owned lot, all of which are provided in a tabular form.

The proposed subdivision must be approved under the Victorian planning scheme. This planning system is developed based on the Planning and Environment Act (State Government of Victoria 1987), which plays a fundamental role in granting a permit for developing and subdividing the land. Therefore, the land surveyor lodges an application for a planning permit with the responsible authority, which is the city council in most cases. City councils mainly assess the planning permit application regarding the following aspects (State Government of Victoria 2012):

- Examination of the proposed subdivision regarding both state and local planning policy frameworks
- The capacity of utility networks and other urban infrastructure to service any new lot to be created as a result of the proposed subdivision
- The possibility of securing access to any new lot
- Provision of open space and other facilities
- The consistency of the proposed subdivision with any formerly approved permit

- The capability of the land to sustain its increased use and development following the subdivision
- The effect and relation of the proposed subdivision on existing buildings located in its neighborhood

According to the above assessment, the city council may undertake the following activities:

- **Referral of planning application:** If the city council decides whether there is any referral required for the proposed subdivision, a copy of the proposal is sent to the relevant referral authorities (State Government of Victoria 2016b). These authorities, such as utility companies, are likely to hold interests over easements and restrictions which are to be created, varied, or removed by the proposed subdivision. In addition, the referral process ensures that responsibilities and assets of referral authorities are not adversely affected by the proposed subdivision. The referral authority should respond to referrals within 28 days after receiving the referrals. If additional information is required by the referral authority, the city council should be informed within 21 days of receipt of the referrals. Once the referral authority is given the additional information, it has, again, 28 days to provide responses to the requested referrals. In responding to referrals, a referral authority advises the city council whether it agrees or disagrees with granting the permit. In some circumstances, when the referral authority concurs with the granting of the permit, it requests the city council to include some conditions in the permit.
- **Request for additional information:** In cases where the city council considers that the information provided within the planning permit application is not sufficient to allow it to make a decision on the planning application, it can request the land surveyor to provide the city council or a referral authority with additional information (State Government of Victoria 1987, 108); (State Government of Victoria 1987).
- **Public notification:** Once the city council is satisfied with the information provided in the planning permit application, it may send a notification to any other party that is detrimentally affected by the grant of the planning permit. In most cases, this includes the owners and residents of lots adjoining the proposed subdivision (State Government of Victoria 1987).

After completing the above activities, the city council will decide whether to grant or refuse the application lodged for a planning permit. If the city council decides to issue a permit, despite having received objections, it will send a Notice of Decision to Grant a Permit to the land surveyor and associated objectors. This notice indicates that the city council has decided to issue the planning permit. The objectors may appeal against granting the permit within 21 days of receiving the notice. The planning permit is usually issued with some conditions. There are typically two types of imposed conditions, namely, those imposed by the responsible authority (the city council) and the ones imposed by referral authorities. The first type of imposed condition is that the city council may oblige the land owner or the surveyor to complete specific actions or provide documents, such as amended plans, for approval. For instance, the city council in most cases must determine an open space requirement, which is typically up to 5 percent of the subdivided land, as a condition of the planning permit (State Government of Victoria 2012, 32). The second type of imposed condition is that the landowner must enter into agreements with utility providers, in which they agree with providing services or further referral of plans before certification. Figure 2.6 provides the BPMN diagram for the planning phase. The details of the information exchanged in this phase are presented in Table 2.1.

2.3.2 Certification Phase

After receiving approval of the planning permit, the land surveyor lodges an application for certifying subdivision plans. It should be noted that the certification process can also be carried out concurrently with the planning permit process (State Government of Victoria 1988, 15). However, the certification of subdivision plans is dependent on the approval of the planning permit. While in the planning phase, the role and functions of the city council are underpinned by the Environment and Planning Act (State Government of Victoria 1987); this authority undertakes the certification process in regard to the Subdivision Act (State Government of Victoria 1988). The subdivision plans provided in the application for certification must also indicate how conditions imposed in the planning permit are met. After a preliminary assessment, the city council may complete the following activities in order to certify subdivision plans:

- **Referral of subdivision plans:** Similar to the planning permit process, the application for certification should also be referred to relevant referral authorities if the city council considers that the proposed subdivision affects their legal interests. The referral authorities review the proposed subdivision according to the conditions they imposed on the planning permit. Responses to referrals must typically be provided to the city council within 28 days of receipt of the application. There could be three decisions made by a referral authority, namely: to refuse consent, to consent to the plan, and/or to require specified alterations.

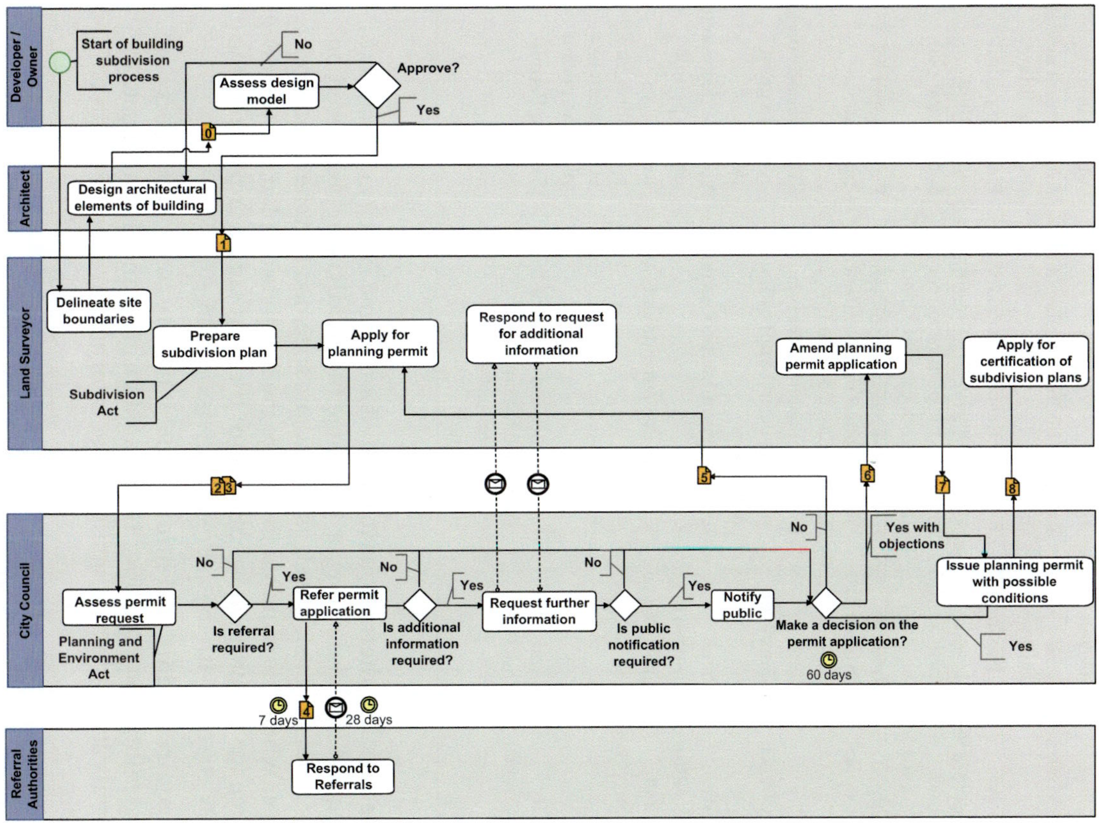

FIGURE 2.6
Planning phase of urban land administration process. (Refer to Table 2.1 for the details of indices in this figure.)

TABLE 2.1

Information Exchanges in Planning Phase

Item No.	Sender	Receiver	Purpose	Content
0	Architect	Owner/ Developer	To approve the design of the building	Architectural model of the building
1	Architect	Land Surveyor	To create subdivision plan of the building	Architectural model of the building
2	Land Surveyor	City Council	To receive the permission for a use or development of land	Planning permit application
3	Land Surveyor	City Council	To receive the permission for a use or development of land	Subdivision plan
4	City Council	Referral Authorities	To receive advice from authorities whose interest may be affected by the grant of a permit	Planning permit application
5	City Council	Land Surveyor	To inform the land surveyor that planning permit application is rejected	Notice of decision to refuse to grant a permit
6	City Council	Land Surveyor	To accept permit with objections	Notice of decision to grant a permit
7	Land Surveyor	City Council	To ensure that amendments are done	Notice of an application for an amendment to a planning permit
8	City Council	Land Surveyor	To allow the land surveyor to subdivide the land and building for use and development	Planning Permit

- **Requesting alternations to subdivision plans:** The city council may request the land surveyor to change the subdivision plan based on its requirements, as well as alterations specified by referral authorities. When it is required to change the subdivision plan, the review time for certification is halted until the land surveyor provides the city council with the altered subdivision plan.
- **Requiring additional information:** If the city council considers that the information provided in the subdivision plan is not adequate for certification, it may request further information about the plan only once.

According to the above activities, the city council makes its final decision on the application for certification, which could be refusal or certification of the subdivision plan. A summary of the certification phase in terms of BPMN notations is provided in Figure 2.7. The details of information exchanged in this phase are presented in Table 2.2.

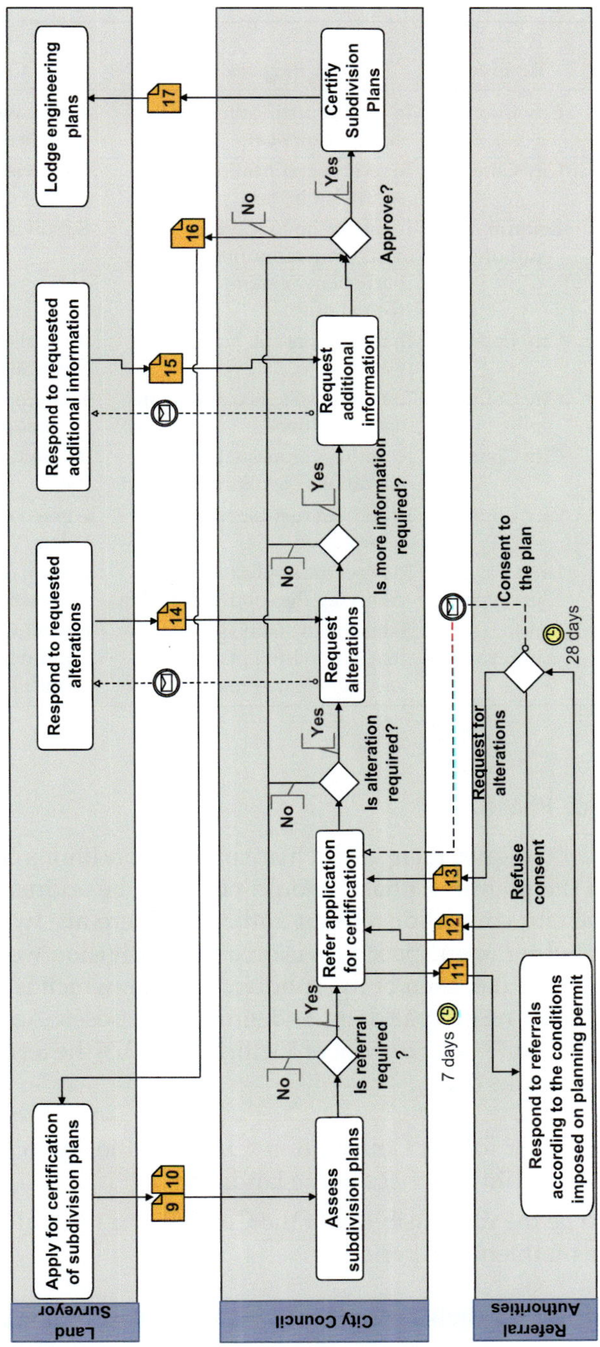

FIGURE 2.7
Certification phase of urban land administration process. (Refer to Table 2.2 for the details of indices in this figure.)

TABLE 2.2

Information Exchanges in Certification Phase

Item No.	Sender	Receiver	Purpose	Content
9	Land Surveyor	City Council	To receive certification for the subdivision plan	Application form for certification of plan
10	Land Surveyor	City Council	To receive certification for the subdivision plan	Subdivision plan
11	City Council	Referral Authorities	To assess the subdivision plans by referral authority, particularly easements and restrictions	Subdivision plan
12	Referral Authorities	City Council	To refuse consent	Refusal of consent by referral authority
13	Referral Authorities	City Council	To require specified alterations and conditions	Requirement for alterations to plan
14	Land surveyor	City Council	To address requested alterations of subdivision plan	Altered subdivision plan
15	Land surveyor	City Council	To address request additional information	Altered subdivision plan
16	City Council	Land Surveyor	To reject the application for certifying the subdivision plan	Refusal of application for certification
17	City Council	Land Surveyor	To inform the land surveyor that subdivision plan is successfully certified	Certification of subdivision plan by council

2.3.3 Compliance Phase

In this phase, several requirements, which are mainly conditions considered on the permit in the planning phase, should normally be addressed to the satisfaction of the city council or referral authority. There are two types of requirements including open space provision and subdivision works. Open space refers to a particular zone of the subdivided land, which is dedicated to public recreation or resort, parkland and similar purposes. According to the Subdivision Act (State Government of Victoria 1988, 32), the term "works" is defined as those works that:

- are required by or for the Council or a referral authority to provide roads or public utility services to the land; and
- are or are to be the responsibility of the Council or a referral authority after the maintenance period.

Prior to commencing subdivision works, the land surveyor should prepare engineering plans in accordance with all specified requirements and imposed conditions in the planning permit. Engineering plans can be lodged with either a referral authority or the city council, or with both, depending on which authority has made the requirements. Upon receipt of engineering plans, a decision must be made within 30 days. Again, there are three possible outcomes determined by the relevant authority, namely, approval, refusal, and a request for alterations. After receiving the approval, the land surveyor can commence subdivision works according to the approved engineering plans.

After completion of the subdivision works, the city council provides the land surveyor with a statement of compliance (SOC) once the following activities are completed:

- The land surveyor has provided the city council with written advice in a prescribed form (State Government of Victoria 2011a, 38). This form advises that all boundaries of roads, reserves, lots, common property areas, and the whole land in the subdivision plan, where appropriate, have been clearly marked out or defined. Additionally, it may notify any substantial discrepancy between subdivision boundaries shown on the certified plan and their counterparts in the real world.
- It is ensured that satisfactory arrangements have been made in accordance with all requirements and conditions placed on the planning permit regarding public works and open space provision. Public works means (State Government of Victoria 1988, 6):
  - The provision of roads, reserves, open spaces, or services within a subdivision
  - Fencing, landscaping, and road works outside the subdivision for roads, reserves, or public open space related to the subdivision
  - Works for sewerage, drainage, water supply, power, gas, or telephone to connect the subdivision to the system serving properties outside it, excluding works to connect any particular property to the system for the subdivision
  - Other prescribed works
- There is an agreement to secure compliance with all requirements mentioned in the previous item.

Figure 2.8 shows the compliance phase in form of a BPMN diagram. The details of information exchanged in this phase are presented in Table 2.3. The next step after obtaining the SOC is applying for registration of the subdivision plans.

2.3.4 Registration Phase

The purpose of this phase is to register subdivision plans. The application for registration is lodged in a land registry organization, which is Land Use Victoria in this case. Besides certified subdivision plans and the issued SOC,

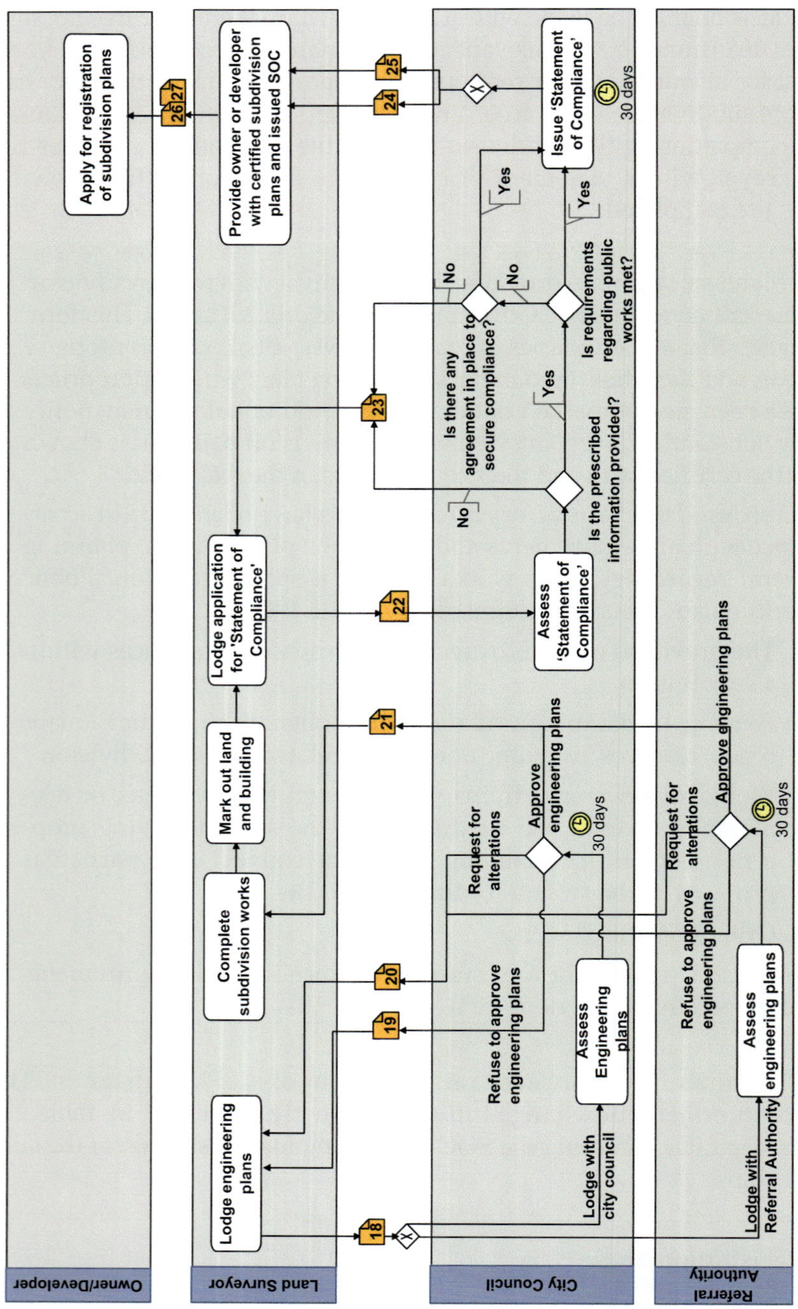

FIGURE 2.8
Compliance phase of urban land administration process. (Refer to Table 2.3 for the details of indices in this figure.)

TABLE 2.3

Information Exchanges in Compliance Phase

Item No.	Sender	Receiver	Purpose	Content
18	Land Surveyor	City Council or Referral Authority	To address all requirements and condition placed on the planning permit	Engineering plans
19	City Council or Referral Authority	Land Surveyor	To refuse to approve the engineering plans	Refusal of engineering plans
20	City Council or Referral Authority	Land Surveyor	To require specified alterations on engineering plans	Requirement of alternations to engineering plans
21	City Council or Referral Authority	Land Surveyor	To confirm that engineering plans are consistent with conditions and requirements of city council and referral authority	Approved engineering plans
22	Land Surveyor	City Council	To notify council that the subdivided land and building have been marked out.	Advice by licensed surveyor
23	City Council	Land Surveyor	To refuse to approve the statement of compliance	Refusal of statement of compliance
24	City Council	Land Surveyor	To ensure compliance with agreements and conditions placed on the planning permit	Statement of Compliance
25	City Council	Land Surveyor	To ensure compliance for specific stage of a staged subdivision	Statement of Compliance for a specified stage of a staged subdivision plan
26	Land surveyor	Owner/ Developer	To provide owner with the certified subdivision plan	Certified subdivision plan
27	Land surveyor	Owner/ Developer	To provide owner with the issued Statement of Compliance	Issued Statement of Compliance

a current certificate of title (Parent Title) for the subdivided land must be provided. In addition, land registry requires the consent of people who are associated with any mortgage, caveat, or encumbrance affecting the subdivided land (Libbis 2015). If subdivision plans include new common property areas, the Owners Corporation information must be additionally provided.

After lodgment, the examiner checks all the required documents. In particular, subdivision plans are rigorously examined in terms of the correctness of legal interests and boundaries. Once the land registry realizes that all requirements associated with registration are satisfactory, it will register the subdivision plan. At this stage, the new lots or pieces of land legally come into existence. The previous legal definition of the subdivided land does not exist anymore, and its title is no longer valid. New title documents are issued for new lots. Additionally, mortgage, caveat, and encumbrance documents, which are the legal documents imposing restrictions on the title documents, may be issued. After registration, the Owners Corporation will undertake responsibilities associated with common property areas. The registration phase is summarized in Figure 2.9. Table 2.4 provides the details of information exchanged in the registration phase.

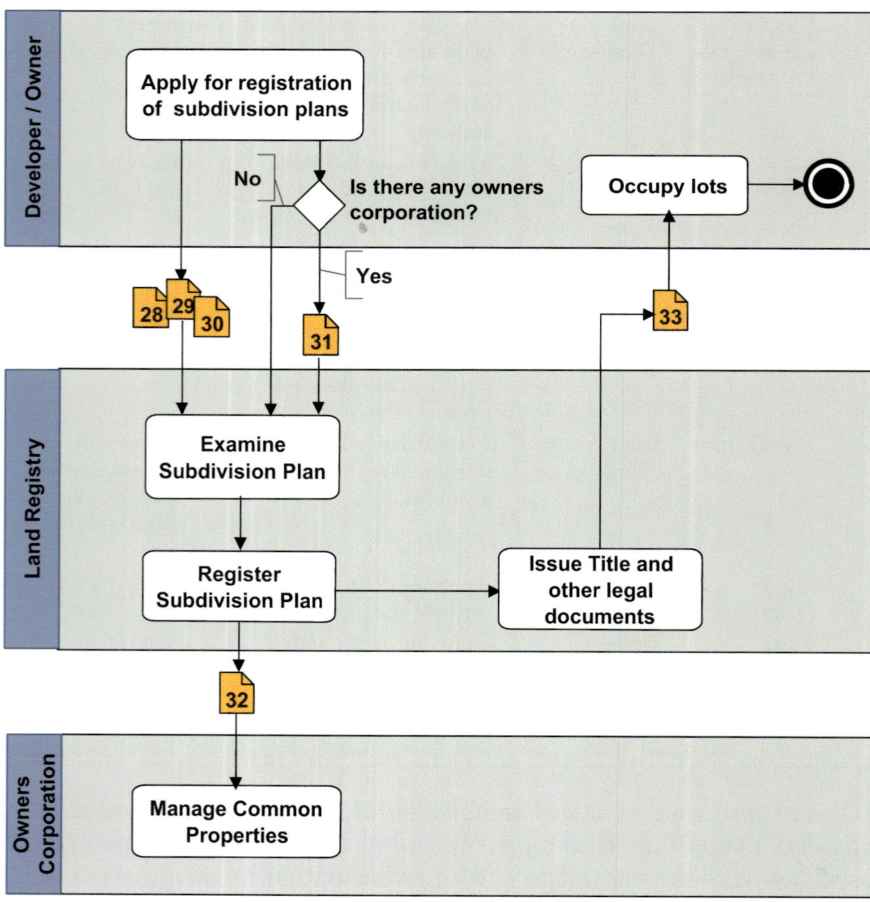

FIGURE 2.9
Registration phase of urban land administration process. (Refer to Table 2.4 for the details of indices in this figure.)

TABLE 2.4

Information Exchanges in Registration Phase

Item No.	Sender	Receiver	Purpose	Content
28	Owner/ Developer	Land Registry	To register the legal interests	Certified subdivision plans
29	Owner/ Developer	Land Registry	To provide land registry with the issued Statement of Compliance	Issued Statement of Compliance
30	Owner/ Developer	Land Registry	To provide land registry with the current title information	Current certificate of title (Parent title)
31	Owner/ Developer	Land registry	To provide land registry with Owners Corporation information	Owners Corporation information
32	Land Registry	Owners Corporation	To manage the common property areas of subdivided development	Registered subdivision plans
33	Land Registry	Owner	To verify registration of legal of lots	New certificate of titles

2.4 Data Requirements in Urban Land Administration

According to the urban land administration process outlined in Section 2.3, five main categories of data requirements are identified. The reason for the importance of these requirements in urban land administration is that they provide the basis for managing complexities associated with legal information defined in urban built environments. These categories include legal objects, physical objects, legal documents, land administration actors, and administrative data. Among these data requirements, legal objects play a core role in urban land administration and data requirements in other categories are to some extent linked to legal objects. Although subdivision plans include legal objects, it is stated in the notation section of these plans the means by which physical objects are used to define the spatial arrangements of legal objects. Therefore, the physical objects category is identified based on the contents of subdivision plans. The legal documents category is identified based on the documents exchanged throughout the building subdivision process. As indicated in Tables 2.1 through 2.5, there are many legal documents exchanged over the process of the building subdivision. However, in this research an important subset of legal documents is selected, a subset of which is critical in each phase of the urban land administration process. Land administration actors mainly refer to those actors who play an important role in performing various activities conducted over the course of an urban development process. The administrative data category is identified as the essential information for certifying and registering subdivision plans.

2.4.1 Legal Objects

There are two main types of legal objects, namely, legal interests and legal boundaries. Legal interests are classified into two categories: primary and secondary. Legal boundaries define the spatial extent of each legal interest. In addition, some attributes are required for managing legal interests. Some attributes are common among all legal interests, while other attributes are required for each specific legal interest. A list of attributes that are common among all legal interests is given in Table 2.5. If there are specific attributes for any type of legal interest, these attributes will be listed in the following subsections with their corresponding legal interest.

2.4.1.1 Primary Legal Interests

Primary legal interests are base level parcels that constitute the continuous cadastral fabric. This means that there is no gap between two primary legal interests, and they cannot overlap each other. These legal interests include lot, common property, road, and reserve.

- **Lot:** A lot can be classified into a land lot and a volumetric lot. A land lot is defined as undeveloped land. This type of lot can be represented by land surface. In the context of urban land administration, volumetric lots are commonly created, and they typically

TABLE 2.5

Required Attributes of All Legal Interests

Attribute	Description
Legal Interest Name	This attribute shows the name of a legal interest. For each legal interest, a specific naming convention is used. For example: Lot: [%]/[Plan Number] e.g. 1/PS123456 Common property: CM [#]/[Plan Number] e.g. CM1/PS123456 Reserve: RES [#]/[Plan Number] e.g. RES1/PS123456
Legal Interest Unit	This attribute describes whether the legal interest consists of one part or multiple parts.
Legal Interest State	This attribute describes the state of legal interest in subdivision plans. Its state could be created, affected, extinguished, and existing.
Legal Interest Class	This attribute refers the specific type of legal interest. Its possible values include lot, common property, road, easement, reserve, crown allotment, crown portion, restriction, depth limitation, and airspace.
Legal Interest Area	This attribute shows the measured area of a legal interest.
Land Use	This optional attribute can be used to specify the land use of a primary legal interest. For instance, land use of a lot could be residential, commercial, and mixed.

include two main parts: an apartment unit and an accessory part (see Figure 2.10). There are two types of accessory part: parking space and storage space. The legal attributes that are specific to lot are given in Table 2.6.

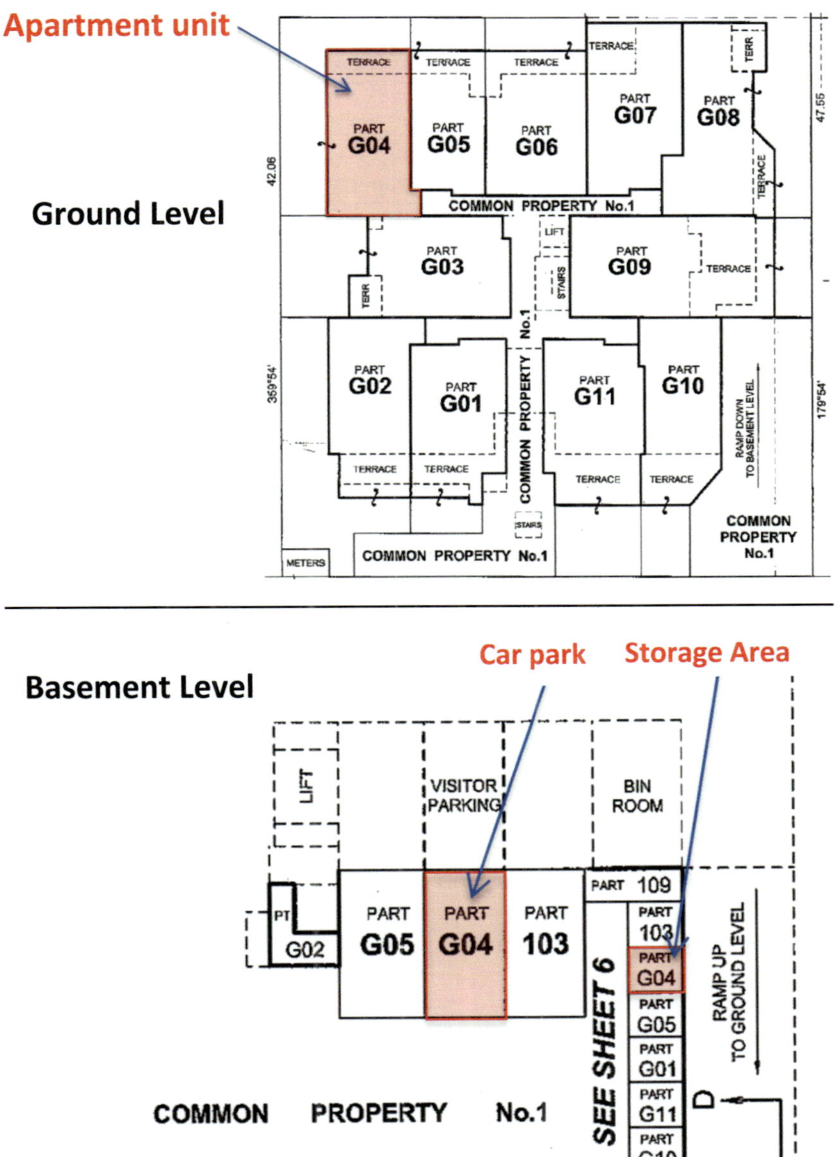

FIGURE 2.10
Parts of a volumetric lot defined within a subdivision plan.

TABLE 2.6

Required Attributes of a Lot

Attribute	Description
Lot Liability	The proportion of the administrative and general expenses that each lot owner is obliged to pay.
Lot Entitlement	The share of interest that each lot owner has in the common property area.
Volume	This attribute shows the measured volume of a volumetric lot.
Address	This attribute should be used to provide the address of lots.

TABLE 2.7

Required Attributes Specific to Common Property

Attribute	Description
Common Property Type	There are two types of common property: limited and unlimited. Unlimited commonly owned properties are for the use and benefit of all owners, whereas limited ones must be used by a specific group of owners.
Land Affected	This includes a list of those lots that are affected by the Owners Corporation being responsible for managing common property.
Limitation Details	This attribute is optional, and it should be used to provide limitation details for limited common properties.

- **Common property:** The common property area of a building is spatially defined as communal legal spaces (for example, corridors and lobbies) plus physical structures (for example, walls and ceilings) that are used by a specific group—or more typically all—of the owners belonging to an Owners Corporation. The legal attributes that are specific to common property are brought in Table 2.7. Common property can be defined by two types, namely, unlimited and limited common property. The former is for the use and benefit of all owners, whereas limited common property must be used by a specific group of owners. As shown in Figure 2.11a, common property No. 1, which is highlighted in red, is unlimited, which means that all lot owners can use and benefit from this common property. However, other common properties are for the use of more specific owners. For example, common property No. 5, which is highlighted in green in Figure 2.11a, is commonly owned by owners of lots with numbers from 101N to 115N. An additional aspect of common properties is that they also include physical structures. 2D cross-section and floorplan views of only communal legal spaces are represented and physical structures are abstracted for the sake of simplicity. In the notation section of the subdivision plan, it is stated that which physical structures belong to the specific common property area. For example, for all lots in Figure 2.11a: All walls defining boundaries,

a)

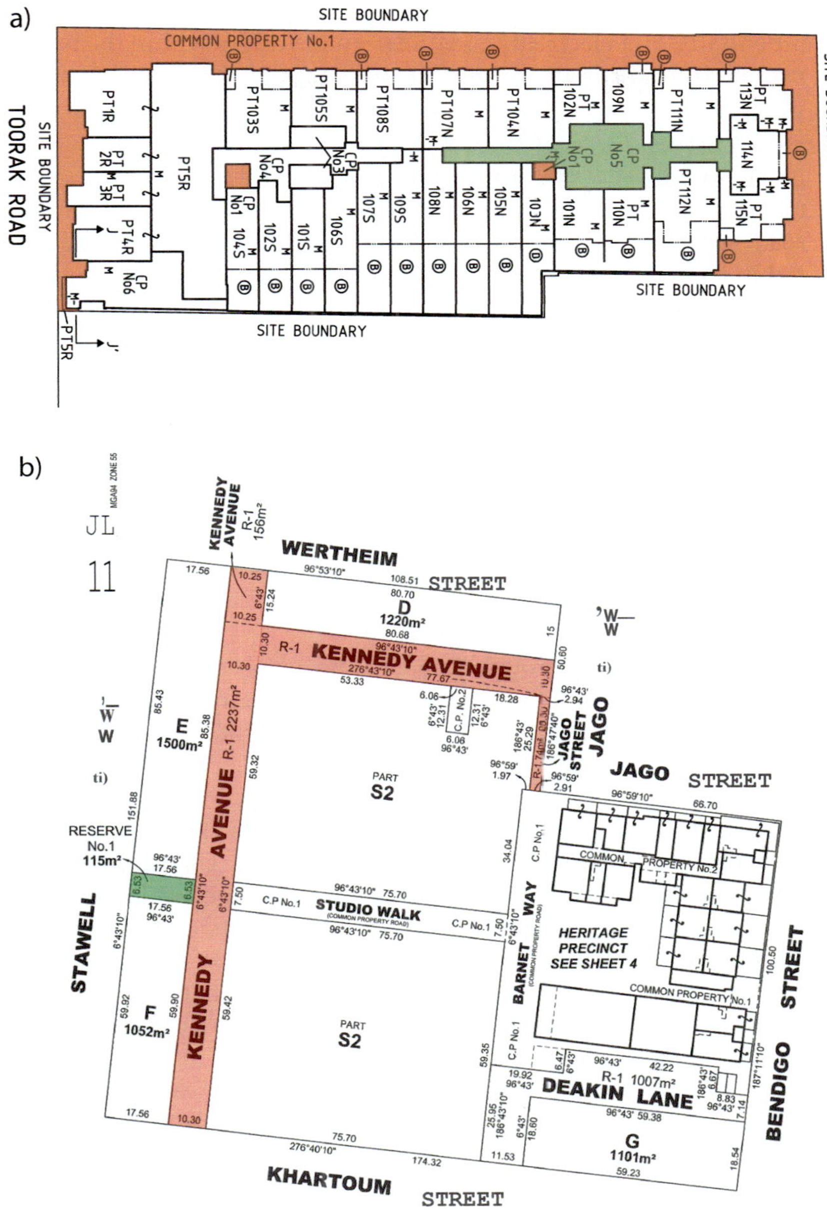

b)

FIGURE 2.11
(a) Unlimited and limited common properties; (b) Examples of road and reserve parcels in subdivision plans.

floor and ceiling slabs, columns, internal service ducts, conduits, pipe shafts, and electricity consumer mains cables within the building and courtyards are deemed to be part of common property No.1.

- **Road:** A road usually includes a carriageway, pavement, verge, and curb. This type of legal interest is for the benefit and use of the public. A complete list of road types is provided in the Victorian ePlan Handbook (State Government of Victoria 2016a). An example of a road legal interest created in a subdivision plan is highlighted in red in Figure 2.11b. The only legal attributes of roads are those ones presented in Table 2.5.
- **Reserve:** A reserve includes those land parcels owned by the city council. An example of this legal interest is highlighted in green in Figure 2.11b. Similar to road, this type of legal interest is for the benefit and use of the public. The City Council usually uses these parcels to establish parks or similar amenities. Reserves are represented as land parcels in subdivision plans. Reserve attributes are only the common ones illustrated in Table 2.5.

2.4.1.2 Secondary Legal Interests

Secondary legal interests provide benefits and/or pose restrictions on primary legal interests. These legal interests can spatially intersect with primary legal interests. They include easement, restriction, depth limitation, and airspace.

- **Easement:** An easement is generally defined as a right to make use of someone else's property without occupying it. This legal interest is typically used for providing a property with a perennial access to essential utility services or right of ways which intersect with neighboring properties. In Victoria, there are two types of easements: private and regulatory. Private easements refer to a right enjoyed by a lot owner over their neighbors' lot. A typical example is a driveway intersecting a neighbor's lot to provide access to a public road. Regulatory easements are held by providers of utilities such as water, electricity, gas, sewerage, and drainage. These utility networks provide usage of, or possibly pose limitations on, the initial land parcel where a building has been developed (see Figure 2.12a). The spatial dimensions of an easement can be defined by either physical (visible) elements, such as utility elements outside the buildings and roads, or it can be defined as invisible external spaces surrounding utility elements. The specific legal attributes of easements are presented in Table 2.8.
- **Restriction:** Restriction is a type of covenant which defines an area on one or more lots where limitations on the use of land apply. For example, as indicated in Figure 2.12b, some lot owners shall not

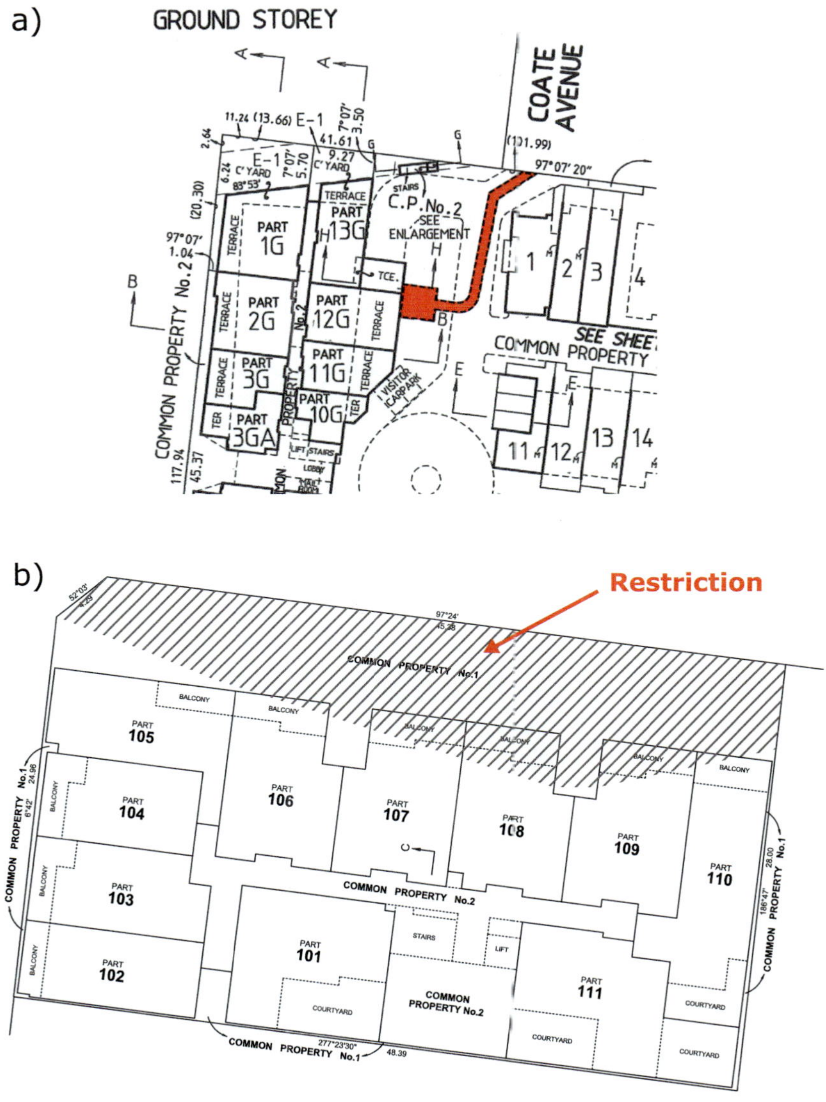

FIGURE 2.12
(a) An easement highlighted in red; (b) An example of a restriction in a subdivision plan.

construct any building requiring sewerage services in a specified area of their land parcels (see Figure 2.12b). Table 2.8 includes those legal attributes which are required for managing restrictions.

- **Depth limitation:** Depth limitation is a form of restriction that originates from the original crown grant. The spatial extent of depth limitation can be determined by underground volumetric spaces

TABLE 2.8
Required Attributes of Easement and Restriction

Legal Interest	Attribute	Description
Easement or Restriction	Land Benefited	This attribute refers to those primary parcels benefiting from the easement or restriction.
	Land Burdened	This attribute refers to those primary parcels on which the easement or restriction impose burden(s).
Easement	Easement Width	The width of the easement measured in length unit (e.g. meter).
	Easement Origin	The number of the plan from which the easement is originated.
	Easement Purpose	Easements can be utilized for a variety of purposes. Examples are drainage, power line, sewerage, floodway, and wetland. The complete list is provided in "Victorian ePlan Handbook, Section 2 –Victoria Reference Data List" (State Government of Victoria 2016a).
Restriction	Restriction Description	Textual description of the restriction.
	Restriction Expiry Date	This attribute is for those restrictions having expiry date.

external to the building. The spatial extent of this legal interest is not delineated in subdivision plans. Only its length is notated in the fact sheet of subdivision plans.

- **Airspace:** Airspace is a form of legal interest that has a spatial extent defined above the ground and external to the building. An example of airspace is height limitation which specifies maximum use of the aboveground space for constructing a building. The spatial extent of this legal interest is not delineated in subdivision plans. Only its length is notated in the fact sheet of subdivision plans. Table 2.9 includes those legal attributes which are specifically required for managing depth limitation or airspace.

2.4.1.3 Legal Boundaries

There is a wide variety of legal boundaries defining spatial extent of legal interests. A taxonomy of these boundaries has been developed by investigating subdivision plans prepared for various types of multi-story building structures. The Subdivision (Registrar's Requirements) Regulations (State Government of Victoria 2011b) underpin the legal basis of defining legal boundaries in Victoria. In the cadastral domain, there is a dichotomy between general and fixed boundaries. We used this dichotomy to develop the taxonomy of legal boundaries in the context of the Victorian jurisdiction.

TABLE 2.9

Specific Legal Attributes of Depth Limitation or Airspace

Attribute	Description
Length	The length of the depth limitation or the height of the airspace, which is typically measured in meters.
Land Applied	Those created lots to which the depth limitation or airspace applies.
Upper Limit	This attribute shows the lower limit of the depth limitation or airspace from the site level.
Lower Limit	This attribute shows the upper limit of the depth limitation or airspace from the site level.

2.4.1.3.1 *General Boundaries*

General boundaries are boundaries that are specified and observed based on real world, tangible, spatial objects. In urban land administration practices, there are three main types of general boundaries: building, ambulatory, and projected (Atazadeh et al. 2017a). Building boundaries are defined and measured by considering the building itself or a part of it. These boundaries can be classified into four categories: interior, median, exterior, and other.

Interior building boundaries are delineated by the interior face of building elements (see Figure 2.13a). This means that all internal fixtures, coverings, and waterproof membranes attached to walls, ceilings, and floors are part of the relevant lot legal interest (State Government of Victoria 2015). For walls, doors, and windows, the internal surface is used to delineate the interior boundary. For floor slabs, the interior boundary is defined by referencing the upper surface, whereas for ceiling slabs, this boundary is defined as the underside of the suspended ceiling.

Median building boundaries lie along the imaginary surface passing through the centerline of building elements. In other words, a median boundary is delineated through referencing a surface having the same distance from both exposed surfaces of a building element. This boundary is delineated to divide the legal ownership of its corresponding building element between the abutting legal spaces. Figure 2.13b represents an example of median boundaries to illustrate how these boundaries are depicted and notated in subdivision plans.

Exterior building boundaries are delineated through referencing the external surface of building elements. These boundaries indicate that the whole of the relevant building elements is contained within the corresponding legal interest. For instance, as illustrated in Figure 2.13c, the boundary between Lot 1 and common property No.1 should be interpreted as the wall, window, and door being completely part of Lot 1.

"Other" building boundaries are delineated through referencing some location, other than the median, inside the building elements. For example, as shown in Figure 2.13d, each level includes a suspended ceiling and

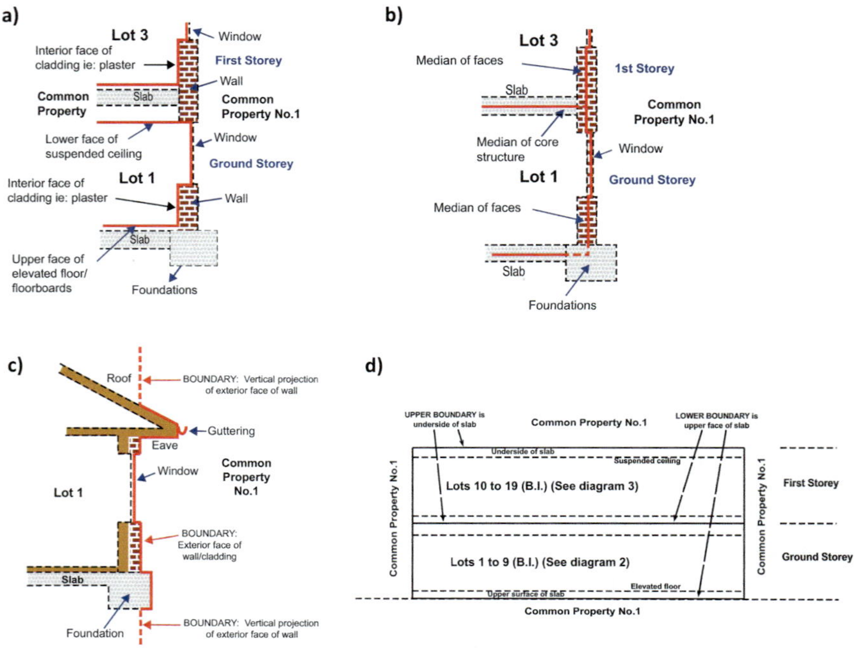

FIGURE 2.13
Building boundaries (a) Interior; (b) Median; (c) Exterior; (d) Other. (Adapted from State Government of Victoria 2015.)

elevated floor and there is a slab between them. As indicated, the upper boundaries touch the underside face of the slab, and the lower boundaries touch the upper face of the slab. This means that both boundaries are inside the whole building element at a location other than the median surface.

As a specific type of general boundary (Williamson et al. 2010, 360), ambulatory boundaries are defined based on observing the movement of dynamic natural features, such as coastlines and river borders (State Government of Victoria 2018). In Figure 2.14a, the floorplan for the ground level of a development comprising four multi-story buildings is represented, and the southern boundary of one of the common properties (Common Property No.1) is delineated by referencing the Yarra river. The location of this boundary may change slowly and imperceptibly over time.

Projected boundaries are mainly used to define balconies and terraced areas of buildings. They are mainly delineated by extending building boundaries in both horizontal, as well as vertical, directions. These boundaries are represented as thick broken lines in cross-section diagrams (see Figure 2.14b). However, in floorplans, projected boundaries are delineated as thick continuous lines. The study showed that it was unusual to extend

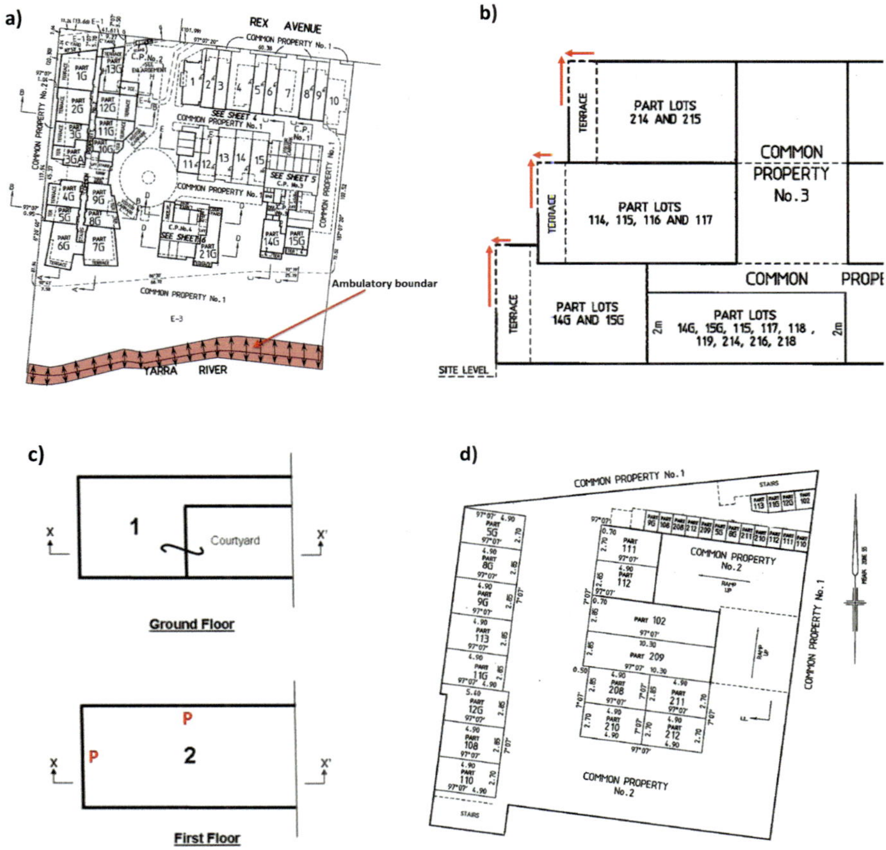

FIGURE 2.14

(a) Ambulatory boundary in a subdivided development; (b) Projected boundaries in cross-section diagrams; (c) Boundaries labeled "P" in the first floor are defined by projection; (d) Fixed boundaries in subdivision plans.

vertical building elements, such as walls, in horizontal directions in any condition. This means that floorplans can only show vertical projections (see Figure 2.14c).

2.4.1.3.2 Fixed Boundaries

Fixed boundaries in subdivision plans are boundaries that are specified based on surveying measurements such as distance, angle, and azimuth. In subdivision plans, this type of boundary is usually utilized to define the spatial extent of parking and storage areas in buildings. In Figure 2.14d, boundaries of a parking area in the basement level of a building are represented, and it can be seen that fixed boundaries are delineated using their corresponding distances and azimuths which are annotated alongside the boundary lines.

TABLE 2.10
Physical Objects

Physical Object	Explanation
Building elements	These physical structures can be used to show the boundaries between legal interests. While walls, windows, doors, and columns provide vertical boundaries, ceilings and floors represent horizontal upper and lower limits of legal interests. Additionally, building elements can constitute parts of common property area.
Distribution elements	Distribution elements inside building can constitute parts of common property areas. Examples are ducts, conduits, pipe shafts, and electricity cables. Distribution elements outside buildings can be defined as easements. Examples are drainage, water supply, and other services.
Geographic elements	These elements can be used to represent geographical easements located around the building site. They can also be used for referencing ambulatory boundaries (e.g. rivers).
Civil elements	These elements can be used to define spatial extent of roads. Therefore, they can be used to represent either part of common property area (which is road) or an individual road legal interest.
Site element	This element represents surface of land, on which the building is developed. It can be used to represent those legal interests whose spatial extent is associated with land surface. These include land lots, reserves, and restrictions.

2.4.2 Physical Objects

Physical objects can enable communication between boundaries referencing physical elements and their respective legal interests. Furthermore, the physical elements can also represent parts of a legal interest. While the spatial extent of these physical objects may not be represented explicitly in subdivision plans, it is shown how these objects define boundaries of legal interests and how they also constitute parts of legal interests (see Table 2.10).

2.4.3 Legal Documents

Legal documents include the legal evidence or facts to specify and certify the interests to be granted for a rightful claimant. As indicated in Section 2.3, there are a large number of legal documents exchanged throughout the urban land administration process. Here, the important ones that play a key role in this process are selected as the main required legal documents. These include: planning permits, certifications, statement of compliance (SOC) documents, engineering plans, title documents, parent titles, mortgage documents, caveats, and encumbrance documents (see Table 2.11). These legal documents are of paramount importance to completing each phase of the urban land administration process.

TABLE 2.11

Important Legal Documents

Legal Document	Explanation
Planning permit	This document provides permission and conditions to use and develop a piece of land.
Certification	This document ensures that the proposed subdivision is satisfactory for the city council.
Statement of Compliance	This document ensures that required activities, such as subdivision works, are undertaken in compliance with agreements and conditions placed on the planning permit.
Engineering plan	If conditions in the planning permit require works to be undertaken (e.g. construction of roads, drainage, and services), engineering plans may be required. In such cases, the statement of compliance will not be issued until the engineering plans have been approved.
Title	Each title document provides an official record of proprietorship of a specific lot. It contains guaranteed information on any person entitled to an interest in the land lot or volumetric lot described on the title document.
Parent title	Parent title provides the legal description of the land before its subdivision.
Mortgage, caveat, and encumbrance	These documents pose restrictions on the title document.

2.4.4 Land Administration Actors

Urban land administration is dependent on a number of actors and their various roles during the urban development process. These land administration actors are involved in the process of subdividing and managing the legal ownership of urban developments. These actors are typically considered as users of legal information. The important actors with descriptions of their main acting roles are provided in Table 2.12. Although a broader range of stakeholders are involved in urban land administration, our view is that the actors listed in Table 2.12 play a fundamental role in making critical decisions in each phase. In Section 2.3, we explained how these actors interact with each other and exchange legal information over the course of the urban land administration process.

2.4.5 Administrative Data

Administrative data are extracted from the fact sheet of subdivision plans. It can be classified into two categories, namely, the survey header and plan information. The survey header contains information regarding the address and geospatial position of the subdivided development, such as parish, township, and approximate geographic coordinates (see Table 2.13). The plan information provides the metadata for the plan of the subdivision, such as plan number, plan area, and plan stage, and the number of lots (see Table 2.14).

TABLE 2.12

Land Administration Actors

Actor	Role Description
Land surveyor	Land surveyor delineates all legal interests in subdivision plans.
Owner	This actor owns a particular lot.
Owners Corporation	Owners Corporation is the legal entity responsible for controlling the use and maintenance of common property areas.
Land Registry	This actor registers all legal interests and issues title documents for lots.
City Council	This authority issues planning permit, certification, and SOC documents. It also owns reserve legal interests.
Referral Authority	This actor mainly owns easements and restrictions. Therefore, it may impose conditions and requirements on the planning permit and specifies whether the easements have been appropriately defined in the subdivision plan. It usually refers to utility companies providing services to buildings.
Developer	This actor has the intention to use and develop a piece of land.

TABLE 2.13

Required Attributes of Survey Header

Attribute	Description
Latitude	It defines the world latitude at the reference point of the subdivided development.
Longitude	It shows the world longitude at the reference point of the subdivided development.
Elevation	Datum elevation at the reference point of the subdivided development relative to the sea level.
Address	Address information about the subdivided development.

TABLE 2.14

Required Attributes of Plan Information

Attribute	Description
Plan Number	Plan number of the subdivided development.
Stage Number	Stage number of the subdivided development.
Lots Number	Number of lots in the subdivided development.
Head of Power	This provides the legislation that the subdivision is based on.
Survey Format	This is the type of survey. Examples are volumetric or building format. Each jurisdiction has its own list of survey types.
Status	This attribute provides the reason for the lodgment of the survey file.
Administrative Date	The date of the survey.

2.5 Data Models Enabling Digital Urban Land Administration

Data models lay the foundation for digitalizing land administration systems. A good data model should be developed in alignment with the data requirements of the land administration process. During the last decade, various types of 3D data models have been developed to support a digital paradigm in urban land administration. Some data models enable digital storage and management of legal dimensions for complex urban environments while others are of general purpose and focus on the physical aspects of built structures. More recently, integrated data models have emerged to harmonize legal arrangements with physical objects in a common data environment. In this section, we will describe some well-known 3D data models enabling digital urban land administration. These include the Land Administration Domain Model (LADM), the Land and Infrastructure (LandInfra) model, the ePlan model, and the legal extensions of CityGML standard.

2.5.1 LADM

The LADM standard defines semantic and geometrical (spatial) data elements required for managing legal information associated with land and properties (ISO19152 2012). Two goals are considered in LADM. Firstly, this standard provides an extendable approach for developing and refining land administration systems. Secondly, LADM attempts to establish a common language to facilitate digital communication of land and property information within one jurisdiction or across various jurisdictions. Three packages, namely Party, Administrative, and Spatial Unit, and one sub-package, Surveying and Spatial Representation, constitute the conceptual schema of LADM (Lemmen, van Oosterom and Bennett 2015, 538).

The Party package includes entities for modeling information about different land administration actors (LA_Party) as well as their roles. These actors could include human or organizational actors. Owner, surveyor, notary, or conveyancer could be examples of roles assigned to actors. The administrative package includes two distinct fundamental concepts, namely, RRR (LA_RRR) and the basic administrative unit (LA_BAUnit). The RRR concept is specialized into "LA_Right," "LA_Restriction," and "LA_Responsibility" which are used for modeling various types of rights, restrictions, and responsibilities respectively. The concept of the basic administrative unit is adopted in LADM to arrange and combine spatial units with the same or homogenous RRR. In other words, RRR information assigned to a basic administrative unit shall be unique (Paulsson and Paasch 2015). For instance, spatial units of an apartment unit, its carparks, and storage areas can be assembled in one basic administrative unit which represents a privately held property right within a multi-level building. The Spatial Unit package and its Surveying and Representation sub-package are used for modeling the spatial dimension of legal objects.

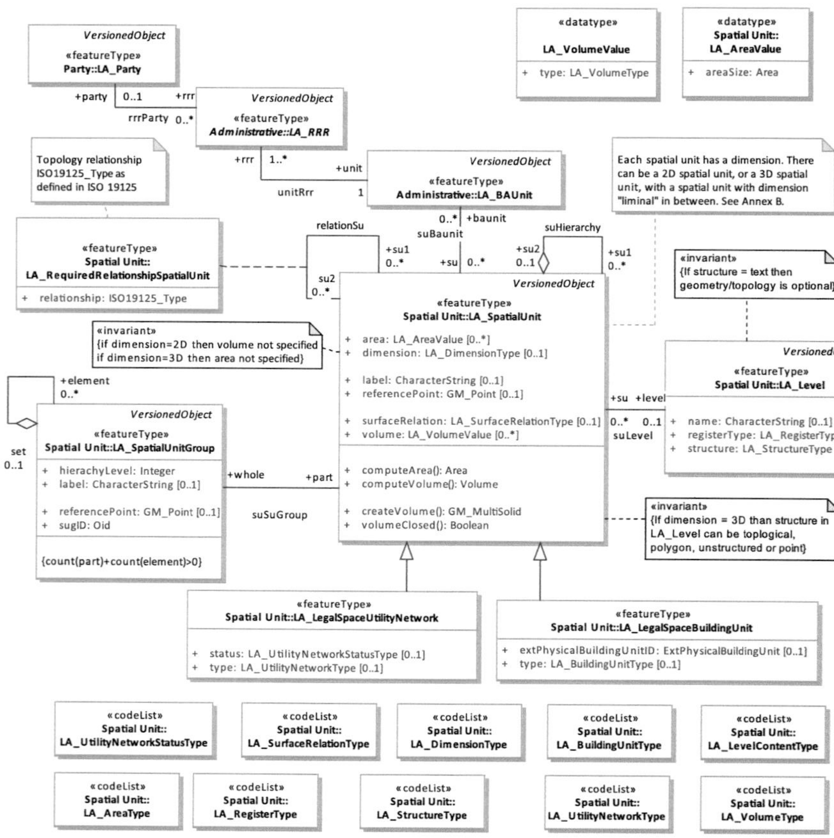

FIGURE 2.15
Detailed entities of Spatial Unit Package. (Adapted from ISO19152 2012, 29.)

As depicted in Figure 2.15, the fundamental entity in the Spatial Unit package is an "LA_SpatialUnit" which is used for modeling the concept of spatial units (Lemmen, van Oosterom and Bennett 2015). This overarching concept includes various representations of the spatial extent of legal interests defined inside any jurisdiction. These representations can be textual descriptions, sketch maps, points, unstructured set of lines, and areal and volumetric legal objects (ISO19152 2012, 82). In the context of 3D RRR data management, there are two specialized concepts of spatial units, namely, building units (LA_LegalSpaceBuildingUnit) and utility networks (LA_LegalSpaceUtilityNetwork). Here a building unit refers to legal space defined for a building or a part of it, which is not essentially equivalent to the physical representation of the building. Likewise, the concept of utility networks in LADM represents legally defined spaces of utilities, which do not essentially equate with their physical counterparts.

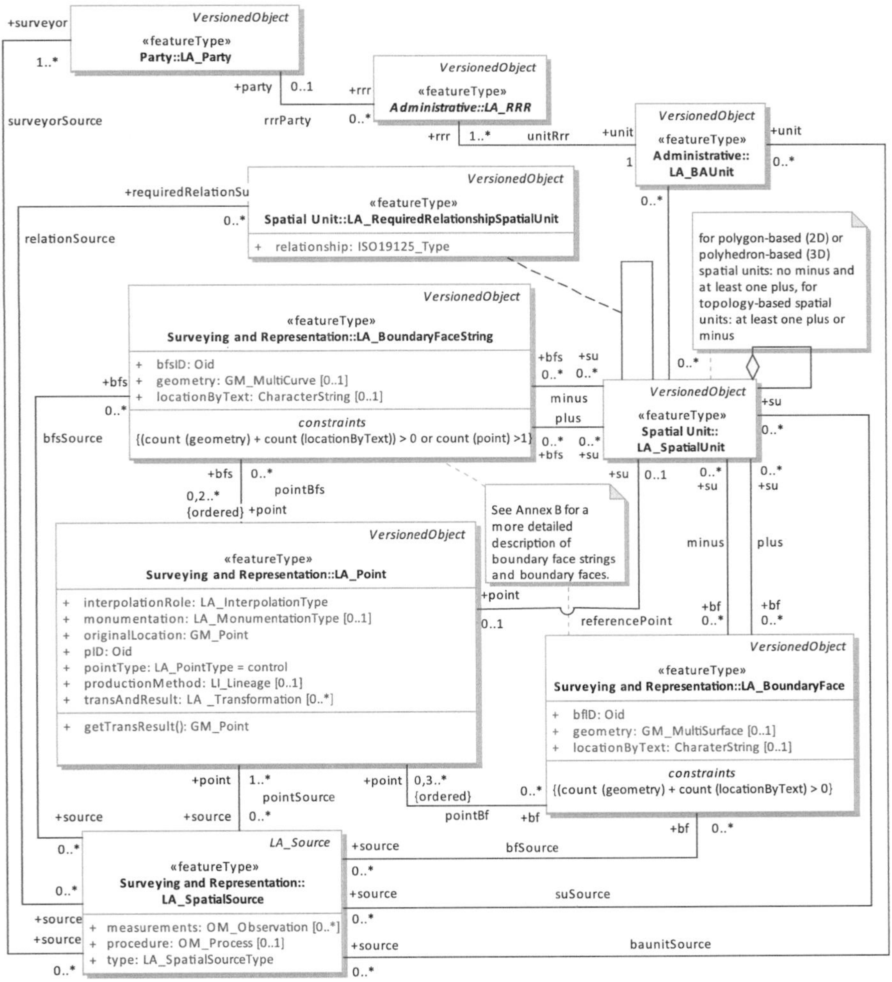

FIGURE 2.16
Basic entities defined in Surveying and Representation sub-package. (Adapted from ISO19152 2012, 37.)

The Surveying and Representation sub-package, which is represented in Figure 2.16, is mainly used to model the topology of boundaries defining spatial units (ISO19152 2012, 32). There are four key entities in this sub package: "LA_Point," "LA_SpatialSource," "LA_BoundaryFaceString," and "LA_BoundaryFace." "LA_Point" is used for modeling points, which would either represent the location of a spatial unit or define a vertex of 2D or 3D boundaries. "LA_SpatialSource" provides information about a set of observations and measurements associated with points. These measurements include distances, azimuths, GPS coordinates, and so on. "LA_BoundaryFaceString" is adopted for modeling geometry of curve boundaries through its association with "LA_Point" and "LA_SpatialSource" entities. Curve boundaries must be defined by at least two points, since the start and end points of a curve are required in the simplest case. "LA_BoundaryFace" is used to represent the geometry of surface boundaries through its association with "LA_Point" and "LA_SpatialSource" entities. A surface boundary must be defined by at least three points. This is because a triangle is the simplest type of surface boundary and three points are required to form a triangle.

With the above introduction of LADM in mind, a few challenges of this legal data model are highlighted in the following.

- First, LADM mainly uses a multi-surface approach, which is not as strong as solid models for modeling 3D spatial objects (Pouliot, Vasseur and Boubehrezh 2013). In the latest version of this legal model, "createVolume()" and volumeClosed()" methods are used to create volumetric spatial units and ensure the closure of the constructed volume, respectively. Although the "createVolume()" method is based on "GM_MultiSolid," there is no entity in the "Surveying and Representation" sub package to define solid models and use them within this method. In addition, "volumeClosed()" only defines Boolean values. However, the Eular–Poincaré formula should be considered as a constraint in this method to ensure closure of volumetric spatial units (see Section 3.4.6.2 in Chapter 3). Therefore, LADM may not adequately allow for solid models to accurately construct the geometric aspect of 3D legal objects. In relation to the management of 3D RRR data, the solid models can facilitate visualization of 3D legal objects and enable the precise volumetric computations and various 3D spatial analyses required in land administration, such as 3D visibility analysis of properties (Jarroush and Even-Tzur 2004; Navratil and Fogliaroni 2014). Implementing such analysis enables the use of 3D RRR data on physical building attributes that may have legal implications, including taxation and valuation.
- The second issue is that LADM do not include sufficient semantic information about 3D properties. In the context of Australian jurisdictions, boundaries of ownership spaces are mainly defined based on physical structures including walls, doors, ceilings, or windows. The "LA_BoundaryFace" entity in LADM does not distinguish a wall boundary surface from a ceiling boundary surface; neither does it specify whether the boundary is located in the interior, exterior face, or median of the physical structure (Atazadeh et al. 2017b). Such ambiguity in defining various boundaries can be resolved by enriching LADM with more semantic information. In addition, sufficient semantic information would eliminate spatial data integration issues and facilitate interoperability of land administration systems with other urban information systems.

- The last issue is that LADM mostly includes legal entities and does not internally incorporate physical information. As mentioned above, in order to form a basis for managing 3D RRR data, LADM uses the concept of spatial units. This concept is able to explicitly represent volumetric legal spaces, but is limited in accommodating situations in which the physical elements are used to delineate the spatial extent of 3D RRR spaces, which is the case in Australia. In response to this issue, LADM provides a limited capability to connect legal objects to their corresponding physical structures via external links. Only the building unit (LA_LegalSpaceBuildingUnit) and the utility network (LA_LegalSpaceUtilityNetwork) classes include *extPhysicalBuildingUnitID* and *extPhysicalUitilityNetworkID* attributes to link legal spaces of building units and utility networks to their physical counterparts, respectively (ISO19152 2012, 29). This indicates that physical information is kept and maintained in other databases with different data structures. In order to integrate two spatial data sets from different sources, two main issues need to be resolved: transforming geometric information and harmonizing semantic information (Isikdag, Underwood and Aouad 2008). Transforming geometric information stems from the fact that the geometry of physical objects may be defined by an approach which is different to the geometry of legal objects. Semantic harmonization refers to alignment of ontology of LADM with physical data models. Even though these geometric and semantic issues are addressed, there might be some discrepancies in matching 3D legal objects to their physical counterparts (Stadler and Kolbe 2007). Therefore, LADM should be internally extended with physical entities.

2.5.2 ePlan Model

Australian jurisdictions have adopted both their national and state ePlan data model based on the LandXML schema (ICSM 2010). LandXML is an implementation XML (eXtensible Markup Language) schema for managing cadastral information obtained from surveying measurements and observations (LandXML 2014).

The "Parcel" class in the ePlan model is used for managing semantic and spatial information about legal interests (see Figure 2.17). Instances of the "Parcel" class can be composed to form a unique ownership interest. The "Parcel" class is specialized into the "VolumetricLot" subclass to semantically define ownership interests with volumetric spatial extent. The "VolumetricLot" entity is associated with the "Volume" entity to define the geometry of 3D RRR spaces (ICSM 2010, 29). A volume in ePlan is a cubic space constituted by polygonal objects, which are its faces. These faces can be both regular and irregular polygons ("Polygon" class) and they define the

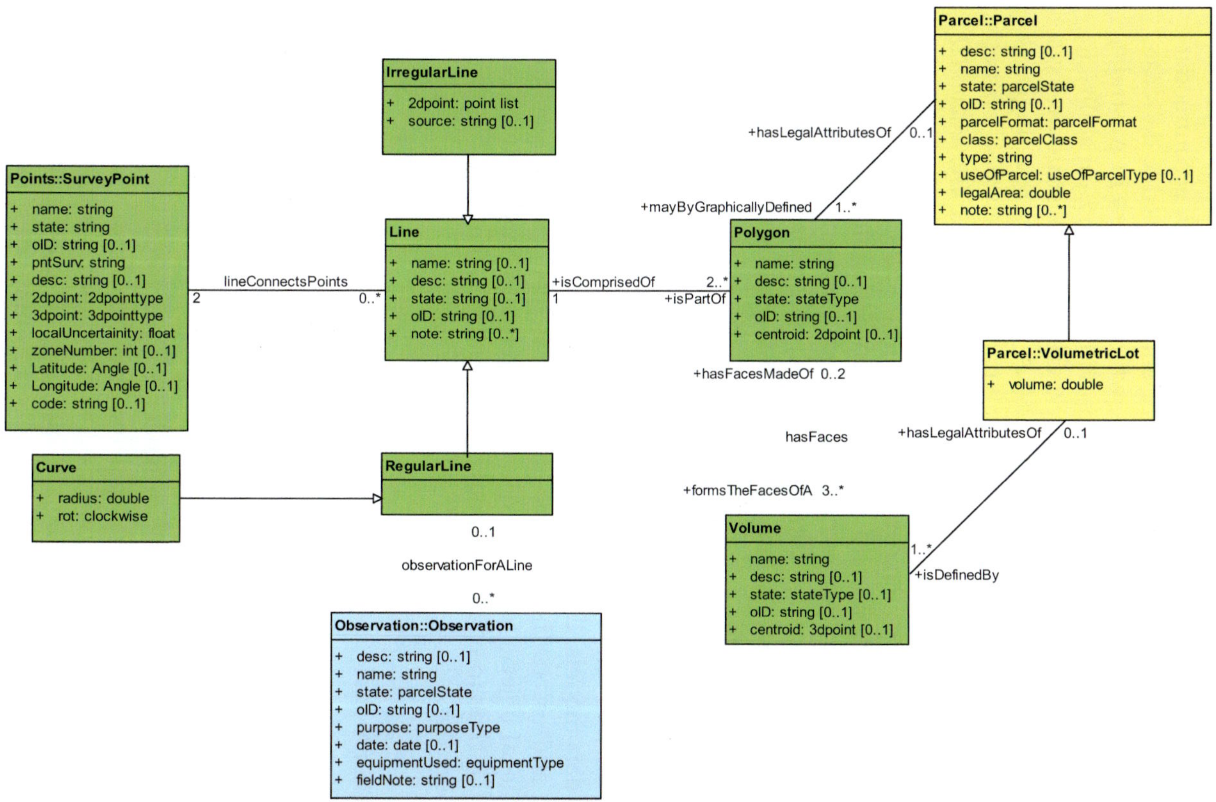

FIGURE 2.17
Parcels and volumetric lots in ePlan model. (Adapted from ICSM 2010, 31.)

boundaries of 3D RRR spaces. A polygon is a closed figure that can be defined by at least two lines. The lines could be regular lines (RegularLine class), curves (Curve Class), or irregular lines (IrregularLine class). "SurveyPoint" and "Observation" classes are used to source the geometry of lines from surveying measurements (see Figure 2.17).

The issues with the ePlan model are similar to those of LADM. First, the ePlan data model does not support solid models for constructing volumetric legal objects. This may cause inconsistencies in terms of the topological closure of volumetric legal objects. Another issue is that the "Polygon" class does not include sufficient semantic information describing boundary types defined based on building elements. There is no attribute in this class to distinguish a wall boundary from a ceiling one. Additionally, an interior boundary cannot be distinguished from median or exterior boundary and vice versa. In ePlan, the only entity for defining external association is "DocFileRef," which is used for referencing external documents. This indicates that ePlan does not even provide any external linkage between legal objects and their physical counterparts. Finally, as depicted in Figure 2.17, the adopted approach for arranging parts of a legal interest utilizes a reflexive relationship (isRelatedTo), which leads to difficulties in semantically differentiating a specific part of a legal interest from its whole spatial structure. For instance, a corridor space, which is part of the common property area, could be incorrectly queried, as the whole common property area in a spatial database is implemented based on the ePlan model.

2.5.3 Land and Infrastructure (LandInfra) Model

The LandInfra conceptual model was developed to model information about land and infrastructure facilities (Scarponcini et al. 2016). The implementation of the LandInfra model is currently under development as the InfraGML encoding standard. This OGC standard comprises a number of requirement classes used for modeling facilities, projects, alignments, roads, rails, surveys, land subdivisions, and wet infrastructure (including storm drainage, wastewater, and water distribution systems). In different parts of this standard, some concepts from LADM and LandXML were adopted for modeling legal objects, while some IFC and CityGML entities were utilized for modeling physical objects. Relevant to modeling ownership interests, there are two requirements classes in LandInfra:

- *LandDivision*: This includes both administrative (which is defined by the "AdministrativeDivision" entity) and cadastral subdivision (ownership interests which are defined by the "InterestsInLand" entity) of land. "PropertyUnit" class, which is defined as a subclass of "InterestsInLand," a generalized entity for modeling an individual unit of ownership. In addition, this entity is equivalent to the concept of basic administrative unit defined in LADM.

- *Condominium*: It models the concurrent ownership interests within buildings, which are mainly private and common properties (see Figure 2.18). Concepts within this requirement class are predicated on the immovable property taxation ADE of CityGML (Çağdaş 2013).

Similar to LADM and LandXML, one issue with LandInfra is that it only models the geometry of legal boundaries, and does not distinguish the semantics of boundary faces. Another issue is that its implementation standard (InfrGML) only supports B-rep solid models, not other types of solid models such as CSG. This standard only considers physical objects (BuildingParts) for defining condominium units. For other ownership interests and land parcels, LandInfra relies on the pure legal concepts of spatial

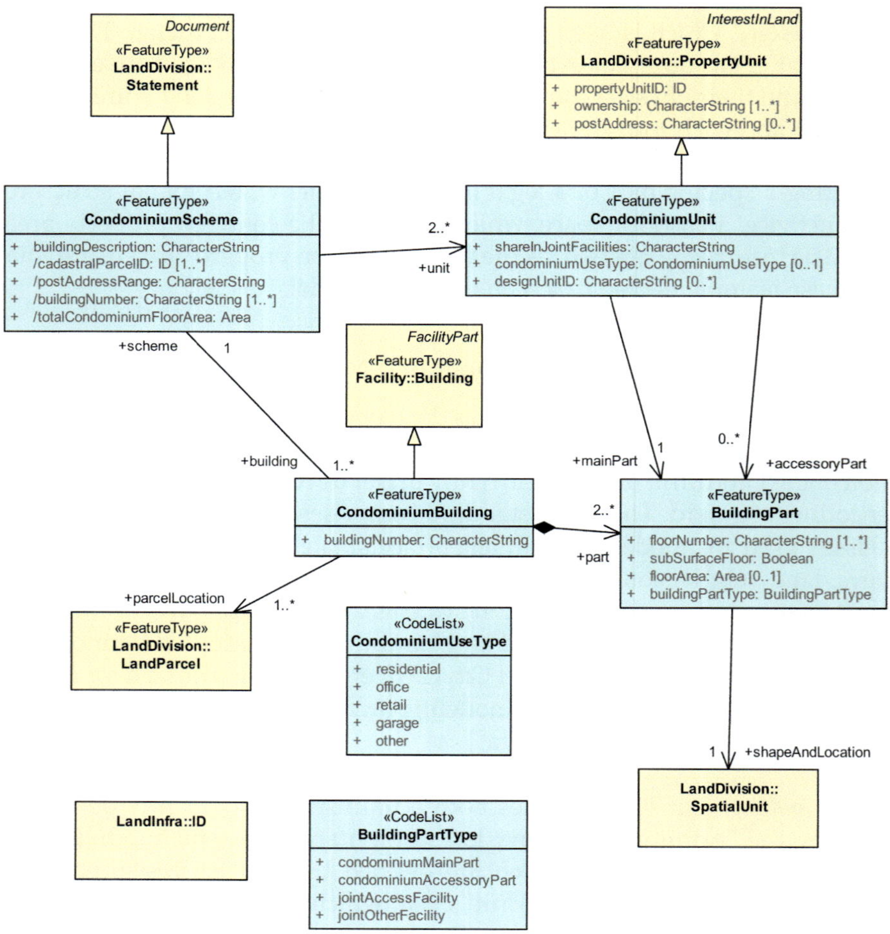

FIGURE 2.18
Condominium context diagram in LandInfra. (Adapted from Scarponcini et al. 2016, 209.)

units acquired from LADM standards. Another interesting thing is that the shape and location of BuildingParts, which are physical concepts acquired from CityGML, are defined based on LADM's spatial unit, which is a legal concept (see Figure 2.18). These issues may stem from the fact that LandInfra is still in the early stages of its development and currently does not have enough capacity to model the spatial complexity inherent in 3D RRR spaces.

2.5.4 CityGML Standard

In the domain of geographic information systems (GIS), CityGML standard is widely known for the interoperable exchange of 3D geospatial data describing the urban built environment (Gröger and Plümer 2012). As an application schema of the Geography Markup Language (GML), CityGML is increasingly the more widely used and implemented model for the 3D representation of urban built environments. This 3D geospatial standard aims at defining a reference ontology for modeling the basic entities, attributes, and spatial relationships required in a 3D urban information model. In other words, this standard defines topological, geometrical, semantical, and appearance properties associated with 3D urban objects. Two fundamental models underpin the spatial data model of CityGML (Kolbe, Gröger and Plümer 2005): a geometric model and thematic model. The geometric model provides both geometrical and topological aspects of 3D urban objects in a consistent and homogeneous framework.

The thematic model assigns the geometric model to various thematic urban objects such as buildings, bridges, vegetation, land use, waterbodies, transportation facilities, and so on. Five Levels-of-Detail (LoD) are distinguished in the CityGML standard to represent 3D urban objects (Gröger and Plümer 2012). Higher LoDs provide more detailed information about 3D city models. The concept of LoD includes both geometric and semantic information. The concept of LoD is represented in Figure 1.9, demonstrating how 3D spatial information about buildings is defined within each LoD. LoD0 provides the least detailed information and only defines 2.5D geometry, in which only one height value is defined for each (x,y) coordinate. LoD1 defines urban objects in terms of generalized 3D models as prismatic block models. LoD2 represents the shape of building roofs and walls, as well as other structures like dormers and balconies. LoD2 also represents objects for vegetation with more detail. For solid waterbodies, LoD2 distinguishes water surface, which is the boundary between water and air, from water ground surface, which is the boundary between water and the ground under water. In LoD3, the outermost shape of urban objects has the highest level of detail. Very detailed 2.5D surfaces are used to show transportation and land use objects. In addition, vegetation shape, waterbodies, and building objects are shown with detailed attributes. For buildings, bridges, and tunnels, openings are also added as thematic objects in LoD3. LoD4 includes interior structural and architectural elements, such as interior walls, furniture, and rooms.

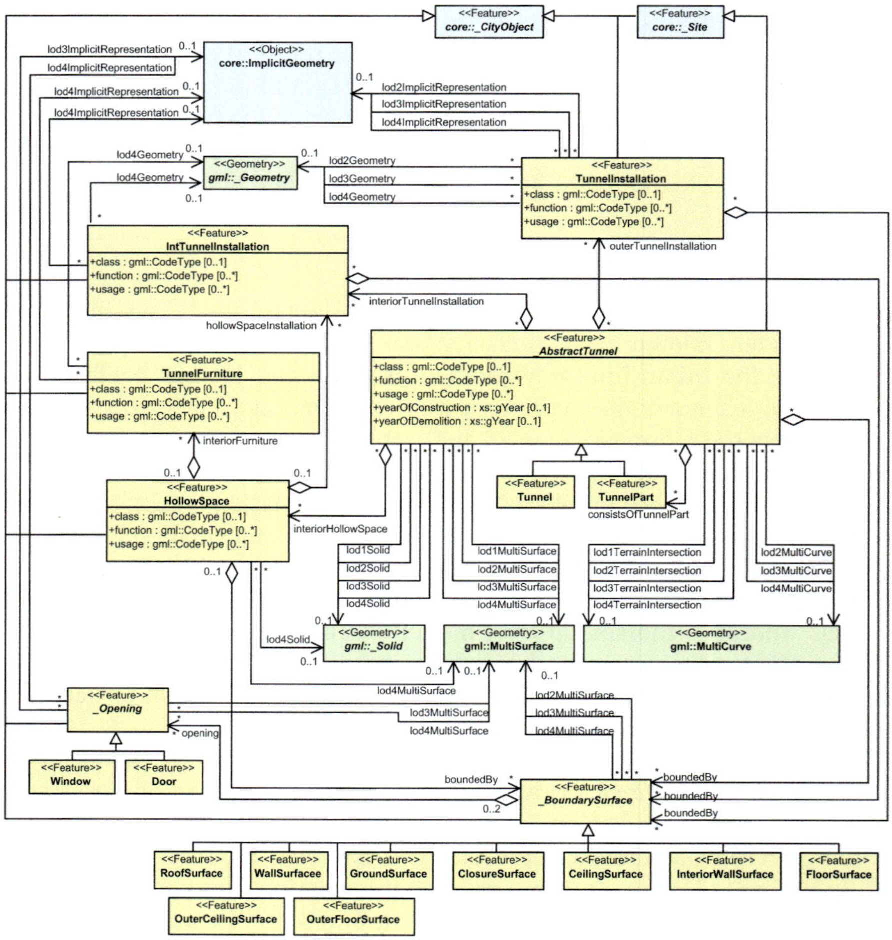

FIGURE 2.19
Building model of CityGML. (Adapted from Groger et al. 2012, 63.)

Relevant to modeling 3D RRR spaces, the geometry of 3D spatial objects in the CityGML standard is developed based on a GML profile. GML, and consequently CityGML, currently implements the boundary representation (Brep) methodology for creating solid models. Brep provides complete information for adjacent topological relationships (Kolbe 2012); however, creating Brep based solid models from the vertexedgeface database is timely and computationally extensive (Nasr and Kamrani 2007). Constructive Solid Geometry (CSG) is a solid modeling method that is more user intuitive, and can enable the construction of complex 3D RRR spaces (Peres and Benhamu 2009); however, CityGML does not support CSG solids at present (Nagel, Stadler and Kolbe 2009).

The finest level of detail (LoD5) in CityGML now provides some data elements for defining the physical boundaries of 3D RRR spaces. The "Room class" is the potential entity for modeling 3D RRR spaces (see Figure 2.19). The corresponding boundaries of the "Room" are determined semantically using the "boundedBy" relationship (Groger et al. 2012). At present, this relationship only extends to interior and exterior wall boundaries (InteriorWallSurface and WallSurface classes), and interior and exterior slab boundaries (GroundSurface, FloorSurface, CeilingSurface, OuterFloorSurface, OuterCeilingSurface, and RoofSurface classes). There are no semantic entities for defining the median wall and slab boundaries, which are often used as a type of building subdivision boundary. There are also no direct semantic relationships which define the "_Opening" class as a boundary of Room class for openings, such as doors and windows. In some cases, columns are used as physical boundaries, particularly in parking areas, between ownership spaces. In CityGML, the "IntBuildingInstallation" class is used for modeling columns within buildings. This class is associated with "Room" class via a "roomInstallation" relationship, which means that an ownership space (an instance of "Room") could possibly contain a column (an instance of "IntBuildingInstallation") or reference another column somewhere else in the building. The implication of this is that the containment or reference is defined, instead of the boundary relationship between ownership spaces and columns.

The core module of CityGML is a physical model of 3D urban objects. Nevertheless, many 3D urban applications necessitate extending the feature types, attributes, and relations for a particular purpose. In light of this, there is an extension mechanism in CityGML, which is known as Application Domain Extension (ADE) (Kolbe 2009). An ADE is an application schema for CityGML and is defined in another XML namespace. New types of features are proposed in each ADE and these features could be subtypes for existing feature types. Moreover, existing feature types can be enriched by incorporating new geometries, attributes, and associations. We will now explain those ADEs developed modeling legal objects inside CityGML.

Rönsdorff, Wilson and Stoter (2014) proposed two options for constructing an LADMbased ADE for the CityGML standard. The first alternative is to create a jurisdiction specific profile of LADM and then implement this profile of LADM as an ADE of CityGML. The second option is to directly implement the fundamental concepts inside LADM, without customizing them for a specific jurisdiction, as an ADE of CityGML. While the latter option was investigated by Rönsdorff, Wilson and Stoter (2014), other investigators have explored the former option in the context of different jurisdictions worldwide (Góźdz et al. 2014; Li et al. 2016).

The general LADM-based ADE proposed by Rönsdorff, Wilson and Stoter (2014) comprises three feature classes, namely "Parcel," "LegalSpace," and "LegalSpaceGroup." These classes were each implemented based on "LA_BAUnit," "LA_SpatialUnit," and "LA_SpatialUnitGroup" concepts in LADM, respectively. To semantically define the RRR information associated with a "Parcel" class, this class was externally linked to "LA_RRR." The advantage of this general integration is the possibility of linking physical and legal objects using an allowed extension mechanism in CityGML without modifying the data structure of LADM. However, the general LADM-based ADE does not provide any details about linking relationships between legal spaces and buildings or building parts.

One jurisdiction-specific implementation of LADM-based ADE of CityGML was conducted by Góździ et al. (2014) within the context of the Polish cadastre (see Figure 2.20). This investigation defined links between the legal and physical spaces of buildings. For managing 3D legal spaces, this ADE proposed "PL_3DParcel" and "PL_LegalSpaceBuilding" classes which are a specialization of the "LA_SpatialUnit" and "LA_LegalSpaceBuildingUnit" (from LADM). For modeling physical objects, two classes were introduced: PL_building (as a subclass of "Building" in CityGML) and PL_BuildingPart (as subclasses of "BuildingPart" in CityGML). In order to link legal objects to their physical counterparts, association relationships between the proposed legal classes and "_AbstractBuilding" class (from CityGML) were also defined within the ADE. This study found that the CityGML standard does not support those elements which are equivalent to non-spatial entities of LADM, such as parties (LA_Party), basic administrate units (LA_BAUnit) and RRR information (LA_RRR). Therefore, these classes were also defined within the proposed integrated model. The feasibility of this LADM-based ADE was tested using three case studies, namely a building located above another development, a residential building located above a public road, and a detached townhouse located on the land parcel. However, all of these implemented case studies were developed in LOD0 and LOD1 of CityGML and other higher LODs (LOD2–4) were not considered in the implementation.

More recently, another LADM-based ADE of the CityGML was developed and implemented by Li et al. (2016) for a Chinese jurisdiction. This ADE includes two separate hierarchies, namely, a legal hierarchy (represented in yellow in Figure 2.21) and a physical hierarchy (represented in light blue in Figure 2.21).

With respect to the spatial structure of condominium units defined in the Chinese jurisdiction, the physical hierarchy comprised three subclasses of "_AbstractBuilding," which are "MajorBody," "Annex," and "SharedObject." "MajorBody" models the representation of the main section of a private ownership space. The legally defined spatial extent of a main body is equivalent to this physical extent. "Annex" entity models those building parts which are used by a specific group of condominium units. This entity also provides information about the gross area, the construction area, and the function of annexes (such as a balcony). "SharedObject" refers to those building parts that are commonly owned and used by all owners of condominium units. Similarly, this entity also provides three attributes, namely, gross area, construction area, and function (such as a staircase or elevator).

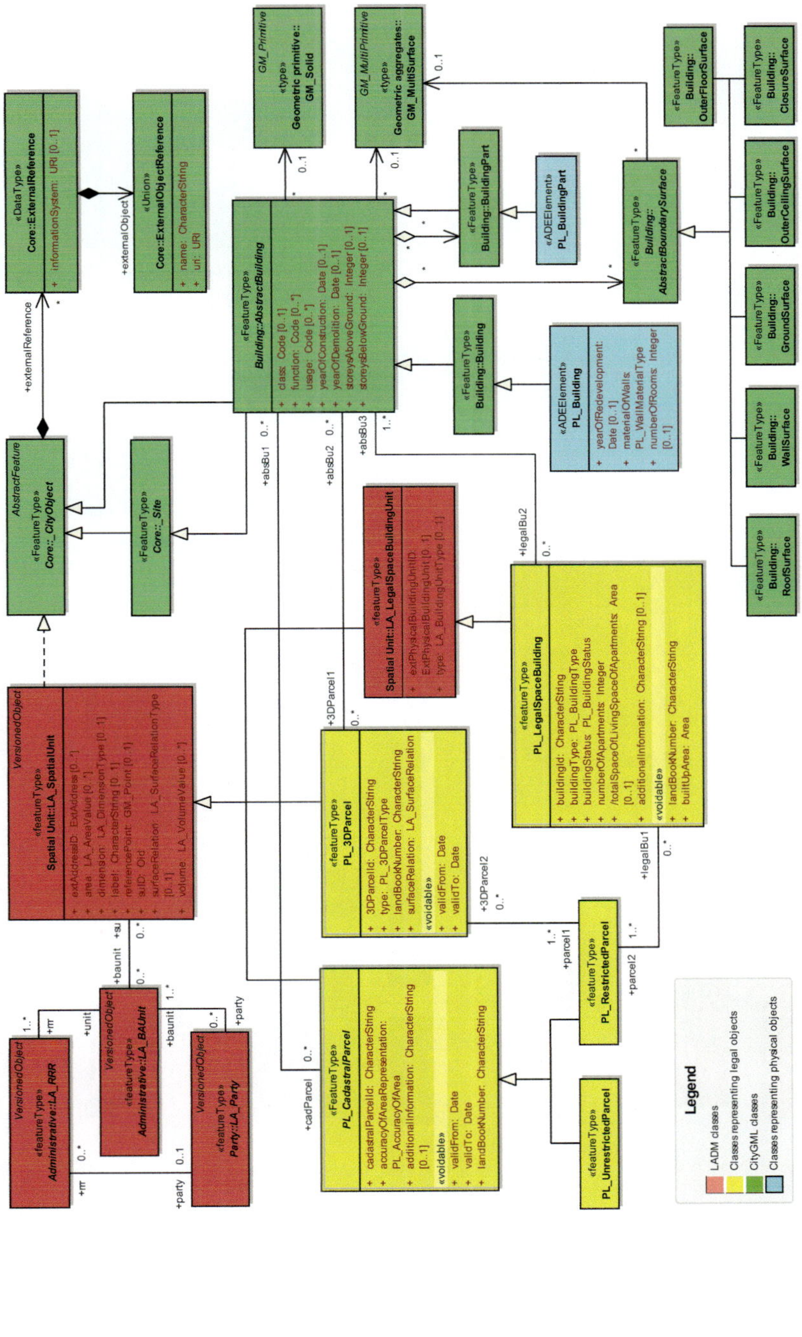

FIGURE 2.20
An LADM-based ADE of CityGML for Polish Cadastre. (Adapted from Góźdź et al. 2014.)

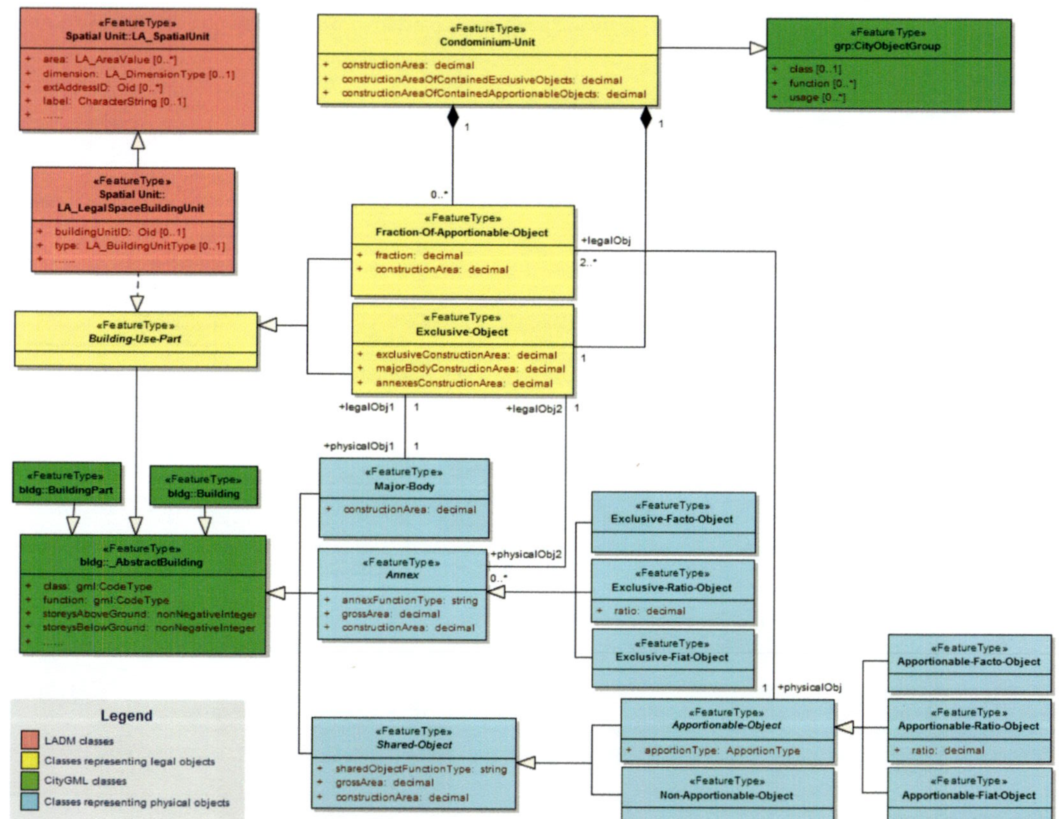

FIGURE 2.21
An LADM-based ADE of CityGML for modeling the spatial structure of condominium units in Chinese jurisdiction. (Adapted from Li et al. 2016.)

The legal hierarchy includes two main classes for modeling legal objects: "Building-Use-Part" and "Condominium-Unit." The former is the subclass of "LA_LegalSpaceBuildingUnit" class within LADM, while the latter is a specialization of the "_CityObjectGroup" class defined inside the thematic extension module "CityObjectGroup" of CityGML. The "BuildingUsePart" was designated for modeling subdivision spaces of a building from an administrative or legal aspect. There are two subclasses of "Building-Use-Part" (see Figure 2.21):

- Exclusive-Object: This subclass models an individually owned property unit as well as associated information, such as an exclusive construction area or a construction area of the major body.
- Fraction-Of-Apportionable-Object: This entity models the virtual elements constituting an apportionable object. There are two attributes within this entity. One attribute is fraction, which provides the proportion of the element to the whole apportionable object of which it is a part. Another attribute describes the construction area of the element.

The "Condominium-Unit" entity models the legal extent of a part of a building, being associated with an individual ownership interest. This class is the aggregation of an "Exclusive-Object" and a "Fraction-Of-Apportionable-Object." This proposed ADE was implemented for two real multi-story buildings, one with a pure residential function and another with mixed commercial and community management functions.

There are two major challenges with the land administration extensions of CityGML:

- Data conversion challenges: The most extensive challenge with the integrated data models reviewed above is that they link CityGML and LADM using connections that are external. The result of this is that data is integrated from two heterogeneous databases, which poses two main challenges: geometric conversion and semantic harmonization (Isikdag, Underwood and Aouad 2008). The geometric conversion comes from the basis that CityGML uses B-rep solids in order to construct 3D physical objects, whereas LADM adopts multi-surface geometry in order to model 3D legal objects. Semantic harmonization means the effective communication of intelligent data from CityGML to LADM or vice versa.
- Insufficiency in defining semantic connectivity between the legal space and physical elements: This challenge indicates that the reviewed integrated data models also did not provide semantic information for one-to-one spatial relationships between physical boundaries and their legal objects. As an example, a one-to-one spatial relationship between a wall and its legal space was not sufficiently modeled.

2.6 Concluding Remarks

In this chapter, the general concept of urban land administration is introduced as the process of recording, managing, and communicating legal information in complex urban built areas. In practice, each jurisdiction considers its own guidelines for implementing land administration systems. Therefore, urban land administration practices in some selected jurisdictions around the globe were overviewed. The common aspect of these practices is their reliance on devising fragmented 2D-based representation methods for managing complex RRR spaces prevailing in urban areas. Upgrading current 2D-based systems to 3D digital environments requires a good understanding of data requirements in urban land administration. Therefore, we have provided a thorough explanation of urban land administration processes in Victoria, Australia, which helped us to identify the main data requirements. This process typically comprises four main phases, namely, the planning phase, the certification phase, the compliance phase, and the registration phase. Through this process, a wide range of data requirements in the context of the Victorian urban land administration system was identified. These requirements were classified into five main categories, including legal objects, physical objects, legal documents, land administration actors, and administrative information. Legal objects play a fundamental role throughout the subdivision process and all other data requirements are associated with this category. For instance, physical objects are required to appropriately record, manage, and communicate the spatial extent of some legal interests. Finally, we introduced various 3D data models which have recently been adopted at an international level for realizing digital paradigms in urban land administration. There are still challenges associated with adopting these data models in delivering a digital urban land administration system. In addressing these challenges, later in this book, we suggest that a BIM environment has more capabilities for responding to the data requirements of urban land administration. In the next chapter, we will introduce the general concepts of BIM and explain the open data model which forms the backbone to a 3D digital, collaborative, and intelligent environment.

References

Andrée, Martin, Jesper Paasch, Jenny Paulsson, and Stefan Seipel. 2018. "BIM and 3D Property Visualisation." In: *FIG Congress 2018*, 6–11 May 2018, Istanbul, Turkey.

Atazadeh Behnam, Mohsen Kalantari, Abbas Rajabifard, and Serene Ho. 2017a. "Modelling Building Ownership Boundaries within BIM Environment: A Case Study in Victoria, Australia." *Computers, Environment and Urban Systems* 61 (Part A): 24–38.

Atazadeh, Behnam, Mohsen Kalantari, Abbas Rajabifard, Serene Ho, and Tuan Ngo. 2017b. "Building Information Modelling for High-Rise Land Administration." *Transactions in GIS* 21 (1):91–113.

Çağdaş, Volkan. 2013. "An Application Domain Extension to CityGML for Immovable Property Taxation: A Turkish Case Study." *International Journal of Applied Earth Observation and Geoinformation* 21 (April):545–555.

El-Mekawy, Mohamed and Anders Östman. 2012. "Feasibility of Building Information Models for 3D Cadastre in Unified City Models." *International Journal of E-Planning Research* 1 (4):35–58.

Góżdz, K., W. Pachelski, P. J. M. Van Oosterom, and V. Coors. 2014. "The Possibilities of Using CityGML for 3D Representation of Buildings in the Cadastre." In: *Proceedings 4th International Workshop on 3D Cadastres*, 339–362, 9–11 November 2014, Dubai, United Arab Emirates. International Federation of Surveyors (FIG).

Groger, G., Thomas Kolbe, Claus Nagel, and K. H. Hafele. 2012. *OGC City Geography Markup Language (CityGML) En-Coding Standard*. Wayland, MA: Open Geospatial Consortium.

Gröger, Gerhard and Lutz Plümer. 2012. "CityGML – Interoperable Semantic 3D City Models." *ISPRS Journal of Photogrammetry and Remote Sensing* 71:12–33.

Ho, Serene. 2014. "Towards 3D-Enabled Urban Land Administration: Invisible Constraints and Strategic Choices." Ph.D. dissertation, University of Melbourne, Australia.

ICSM. 2010. "EPlan Model." The Intergovernmental Committee on Surveying and Mapping, Canberra, Australia. Available: www.icsm.gov.au/sites/default/files/ICSM-ePlan-Model-v1.0 %281%29.pdf.

Isikdag, Umit, Jason Underwood, and Ghassan Aouad. 2008. "An Investigation into the Applicability of Building Information Models in Geospatial Environment in Support of Site Selection and Fire Response Management Processes." *Advanced Engineering Informatics* 22 (4):504–519.

ISO19152. 2012. *Geographic Information––Land Administration Domain Model (LADM)*. Geneva, Switzerland: International Organization for Standardization (ISO).

Jarroush, Jad and Gilad Even-Tzur. 2004. "Constructive Solid Geometry as the Basis of 3D Future Cadastre." In: *FIG Working Week*, 22–27 May 2004, Athens, Greece.

Kitsakis, Dimitrios, Jesper Paasch, Jenny Paulsson, Gerhard Navratil, Nikola Vučić, Marcin Karabin, Mohamed El-Mekawy, Mila Koeva, Karel Janečka, and Diego Erba. 2018. "Chapter 1: Legal Foundations." In: Peter van Oosterom (Ed.), *Best Practices 3D Cadastres*, 1–66, 6–11 May 2018, Istanbul, Turkey. International Federation of Surveyors (FIG).

Kolbe, Thomas. 2009. "Representing and Exchanging 3D City Models with CityGML." In: J. Lee and S. Zlatanova (Eds.), *3D Geo-Information Sciences*, 15–31. Berlin/Heidelberg: Springer.

Kolbe, Thomas. 2012. "BIM, CityGML, and Related Standardization." In: *Digital Landscape Architecture Conference 2012*, 31 May–2 June 2012, Bernburg/Dessau, Germany.

Kolbe, Thomas, Gerhard Gröger, and Lutz Plümer. 2005. "CityGML: Interoperable Access to 3D City Models." In: Peter van Oosterom, Siyka Zlatanova, and Elfriede M. Fendel (Eds.), *Geo-Information for Disaster Management SE - 63*, 883–899. Springer.

LandXML. 2014. "LandXML 2.0 (Working Draft) Schema Announced." Available at: http://landxml.org/.

Larsson, Karolina, Jesper Paasch, and Jenny Paulsson. 2018. "Conversion of 2D Analogue Cadastral Boundary Plans into 3D Digital Information: Problems and Challenges Illustrated by a Swedish Case." In: *6th International FIG 3D Cadastre Workshop*, 75–94, 2–4 October 2018, Delft, the Netherlands.

Lemmen, Christiaan, Peter van Oosterom, and Rohan Bennett. 2015. "The Land Administration Domain Model." *Land Use Policy* 49:535–545.

Li, Lin, Jindi Wu, Haihong Zhu, Xinqiao Duan, and Feng Luo. 2016. "3D Modeling of the Ownership Structure of Condominium Units." *Computers, Environment and Urban Systems* 59 (September):50–63.

Libbis, Simon. 2015. *Subdivisions Victoria: The Ultimate Guide*. Hybrid Publishers.

Nagel, Claus, Alexandra Stadler, and Thomas Kolbe. 2009. "Conceptual Requirements for the Automatic Reconstruction of Building Information Models from Uninterpreted 3D Models." In: *Proceedings of the International Archives of Photogrammetry, Remote Sensing and Spatial Information Sciences*, 34 (XXX):46–53.

Nasr, Emad Abouel and Ali K. Kamrani. 2007. "Methodologies of Feature Representations." In: *Computer-Based Design and Manufacturing*, 71–100. Springer US. doi:10.1007/978-0-387-23324-6_4.

Navratil, Gerhard and Paolo Fogliaroni. 2014. "Visibility Analysis in a 3D Cadastre." In: Peter van Oosterom, and Elfriede Fendel (Eds.), *Proceedings of the 4th International Workshop on 3D Cadastres*, 183–196, 9–11 November 2014, Dubai, United Arab Emirates. International Federation of Surveyors (FIG).

Olfat, Hamed, Davood Shojaei, Mark Briffa, Susannah Maley, and Abbas Rajabifard. 2018. "Strategic Actions for Increasing the Submission of Digital Cadastral Data by the Surveying Industry Based on Lessons Learned from Victoria, Australia." *ISPRS International Journal of Geo-Information* 7 (2): 47.

Paulsson, Jenny. 2007. *3D Property Rights–An Analysis of Key Factors Based on International Experience*. KTH Royal Institute of Technology.

Paulsson, Jenny and Jesper M. Paasch. 2015. "The Land Administration Domain Model – A Literature Survey." *Land Use Policy* 49 (December):546–551.

Peres, Nurit and Moshe Benhamu. 2009. "3D Cadastre – Geometry, Topology and Other Technical Considerations." In: *FIG Working Week*, 1–14, 3–8 May 2009, Eilat, Israel.

Pouliot, Jacynthe, R. Cuissart, and P. Bordin. 2015. "Cadastral Mapping for Underground Networks: A Preliminary Analysis of User Needs." In: *27th International Cartographic Conference*, 23–28 August 2015, Rio de Janeiro, Brazil. International Cartographic Association.

Pouliot, Jacynthe, Tania Roy, Guillaume Fouquet-Asselin, and Joanie Desgroseilliers. 2011. "3D Cadastre in the Province of Quebec: A First Experiment for the Construction of a Volumetric Representation." In: Thomas Kolbe (Ed.), *Advances in 3D Geo-Information Sciences*, 149–162. Springer.

Pouliot, Jacynthe, Marc Vasseur, and Abbas Boubehrezh. 2013. "How the ISO 19152 Land Administration Domain Model Performs in the Comparison of Cadastral Systems: A Case Study of Condominium/Co-Ownership in Quebec (Canada) and Alsace Moselle (France)." *Computers, Environment and Urban Systems* 40:68–78.

Rajabifard, Abbas, Muyiwa Agunbiade, Mohsen Kalantari, Kit Meng Yip, Behnam Atazadeh, Farshad Badiee, Dato' Mohd Noor Bin Isa, et al. 2018. "An LADM-Based Approach for Developing and Implementing a National 3D Cadastre – A Case Study of Malaysia." In: Christiaan Lemmen, and Peter van Oosterom

(Eds.), *7th International FIG Workshop on the Land Administration Domain Model*, 47–66, 12–13 April 2018, Zagreb, Croatia. Available at: https://wiki.tudelft.nl/pub/Research/ISO19152/LADM2018Workshop/04-18_LADM_2018.pdf.

Rönsdorff, C., D. Wilson, and Jantien Stoter. 2014. "Integration of Land Administration Domain Model with CityGML for 3D Cadastre." In: *Proceedings of 4th International Workshop on 3D Cadastres*, 313–322, 9–11 November 2014, Dubai, United Arab Emirates. International Federation of Surveyors (FIG).

Scarponcini, Paul, Hans-Christoph Gruler, Erik Stubkjær, Peter Axelsson, and Lars Wikstrom. 2016. "OGC® Land and Infrastructure Conceptual Model Standard (LandInfra)."

Stadler, Alexandra and Thomas Kolbe. 2007. "Spatio-Semantic Coherence in the Integration of 3D City Models." In: *Proceedings of the 5th International Symposium on Spatial Data Quality*, 13–15 June 2007, Enschede, the Netherlands.

State Government of Victoria. 1987. *Planning and Environment Act 1987.* Victoria, Australia. Available at: http://classic.austlii.edu.au/au/legis/vic/consol_act/paea1987254/.

State Government of Victoria. 1988. *Subdivision Act 1988.* Victoria, Australia. Available at: http://www8.austlii.edu.au/cgi-bin/viewdb/au/legis/vic/consol_act/sa1988153/.

State Government of Victoria. 2011a. *Subdivision (Procedures) Regulations 2011.* Victoria, Australia. Available at: www8.austlii.edu.au/cgi-bin/viewdb/au/legis/vic/num_reg/sr2011n112o2011457/.

State Government of Victoria. 2011b. *Subdivision (Registrar's Requirements) Regulations 2011.* Victoria, Australia. Available at: www.austlii.edu.au/cgi-bin/download.cgi/cgi-bin/download.cgi/download/au/legis/vic/consol_reg/srr2011538.pdf.

State Government of Victoria. 2012. *Subdivision Act User Guide.* Melbourne, Australia.

State Government of Victoria. 2015. *Building Subdivision Guidelines.* Melbourne, Australia. Available at: www.propertyandlandtitles.vic.gov.au/__data/assets/word_doc/0027/53946/Building-subdivision-guidelines-update-delwp.doc.

State Government of Victoria. 2016a. *EPlan Handbook.* Melbourne, Australia. Available at: www.spear.land.vic.gov.au/spear/documents/eplan/ePlan Handbook 1.7.pdf.

State Government of Victoria. 2016b. *Using Victoria's Planning System.* Melbourne, Australia. Available at: www.planning.vic.gov.au/__data/assets/pdf_file/0017/95012/Using-Victorias-Planning-System-2015.pdf.

State Government of Victoria. 2018. *Ambulatory Boundaries.* Department of Environment, Land, Water and Planning. Available at: www.dtpli.vic.gov.au/property-and-land-titles/surveying/advice-and-guidelines-for-surveyors/ambulatory-boundaries.

Stoter, Jantien. 2004. "3D Cadastre." Technische Universiteit Delft.

Stoter, Jantien, Hendrik Ploeger, and Peter van Oosterom. 2013. "3D Cadastre in the Netherlands: Developments and International Applicability." *Computers, Environment and Urban Systems* 40 (July):56–67.

White, Stephen A. and Conrad Bock. 2011. *BPMN 2.0 Handbook: Methods, Concepts, Case Studies and Standards in Business Process Management Notation.* 2nd edn. Future Strategies Inc.

Williamson, Ian P., Stig Enemark, Jude Wallace, and Abbas Rajabifard. 2010. *Land Administration for Sustainable Development.* Redlands, CA: ESRI Press Academic.

3

Fundamentals of the BIM Environment—Opportunities for Land Administration

3.1 Introduction

The long-established practice within the architecture, engineering, and construction (AEC) industry has mainly been predicated on using document-driven methods to store, manage, and exchange geometric and semantic information during the development process of a built asset. The documents exchanged among AEC actors typically include 2D-based and specialized methods of representing building information. These documents are not easily understood by other stakeholders with limited technical backgrounds. Inadequate communication of building information using 2D-based approaches often leads to errors, field costs, slow project delivery, and legal disputes among the stakeholders and organizations involved in a construction project. The traditional practice in the AEC industry has led to fragmentation in terms of unexpected project costs, delays, legal issues, and so forth. Therefore, more efficient approaches should be considered to facilitate communication and collaboration in AEC projects.

Advancements in techniques for 3D digital modeling have accelerated the production of 3D physical models of buildings and other urban infrastructure worldwide (Becker, Nagel and Kolbe 2013). These 3D models have proliferated in different formats and levels of detail (Hammoudi 2011; Kolbe, König, and Nagel 2011). Although the use of 3D geometric models in the AEC industry has reduced the amount of time taken by the exchange of building information, these models have had an insignificant impact on reducing conflicts stemming from 2D documents. Among 3D building models, the most comprehensive and intelligent 3D digital approach to managing information during the lifespan of complex and multi-level buildings is Building Information Modeling (BIM). BIM has shifted the AEC industry from image-based or geometric representation methods towards intelligent data management approaches. This paradigm provides the pathway from silo-based data exchange methods towards an integrated, interoperable, common, and coordinated data-sharing environment.

A BIM environment has been developed to facilitate communication and collaboration over the course of a building lifecycle (Singh, Gu and Wang 2011). Since BIM improves interorganizational communication, it has a considerable influence on enhancing productivity and efficiency, and on cost reduction in the planning, design, construction, operation, and management phases during the development process of AEC projects. BIM tools can provide various views of BIM models in terms of 2D floorplans or cross-section diagrams, as well as 3D models. These tools also provide optimal methods for spatial representation, simulation of building behavior, and efficient collaboration among AEC actors. Interoperability of BIM tools plays a fundamental role in enabling full collaboration among the users of these tools. As a result, the BuildingSMART organization has been developing various open standards over the last decade to encourage AEC organizations to collaborate in open BIM environments, with the Industry Foundation Classes (IFC) being one of the major standards.

Since BIM is a very new topic in the domain of land administration, some readers may not be familiar with its concepts. Therefore, the aim of this chapter is to provide a general introduction to the fundamental concepts behind the paradigm of BIM and to elaborate its importance in the land administration domain. Moreover, open BIM standards and relevant IFC entities for urban land administration are described. A comparison between the IFC standard and other 3D spatial data models, particularly the CityGML standard, will also be presented in this chapter to highlight the capabilities of the IFC standard in digitizing urban land administration data. In the last section of this chapter, an overview of relevant BIM projects in the land administration domain will be presented.

3.2 BIM

The foundation of the BIM-based paradigm is underpinned by two related computer-based models, namely Computer Aided Design (CAD) models, and building product models (Isikdag 2015, 1). CAD models comprise a purely geometric representation of building elements in either 2D or 3D. This implies that these models only include sets of polylines and polygons to describe building elements. There is no semantic, functional, or intelligent information about building elements stored in CAD models. In response, building product models have been developed to manage both semantic and geometric information describing building products (Eastman 1999). These models utilized STEP (STandard for Exchange of Product model data) as a formal standard to enable sharing product data (Pratt 2001).

Over the last decade, there have been significant developments in computer graphics in terms of rendering and manipulating 3D models. Consequently, there has been a considerable trend towards leveraging building product models, resulting in the appearance of the BIM-based paradigm in the AEC and facility management industries. BIM offers more intelligence and interoperability compared to traditional 3D CAD models. Digital management of both physical and functional characteristics of buildings inside the BIM environment facilitates multi-disciplinary team work in AEC projects. Since BIM provides the ability to store building information in a structured and logical way, any modification in the design data can be clearly handled over the course of a project lifecycle. The BIM environment provides comprehensive datasets describing the geometry and semantics of building components. In addition, a wide range of connectivity relationships between building components are stored in BIM models. BIM models often include geometric information, energy analysis, spatial relationships, geospatial information, quantities and properties of building elements, invisible functional spaces, material of building elements, cost information, structural analysis, and so on.

The BIM environment plays a significant role in addressing the spatial and functional complexities of various dimensions in multi-story developments. The integrated approach offered in this environment enables stakeholders to efficiently share spatial and semantic information about complex building structures; for instance, an architectural BIM model of a building can be better understood by a structural engineer. BIM provides considerable economic benefits. Return on investment (ROI) is an important factor considered in measuring economic benefits resulting from the adoption of BIM in AEC firms around the world (Giel 2009). Most surveys report high ROI values for adopting BIM in both short and long periods of time (Lee, Park and Won 2012). One significant short term benefit of the BIM environment is decreasing errors and omissions to the smallest possible amount. Another short term benefit is less turnover of staff members in AEC firms. Reduction in construction costs and fewer contractual claims are also considered as major long term benefits stemming from the BIM-based paradigm adopted in the AEC industry.

The BIM environment can be considered as an efficient decision support system to achieve the objectives of a project in an optimal way (Kim et al. 2015a). The 3D digital data-sharing space of this environment enables the project team to persistently evaluate and monitor building information. Spatially complex information can be verified visually within the BIM environment, which may subsequently increase the satisfaction level of the project client. Visual representation of building information reduces the amount of time required for exchanging complex ideas among the project team. In the BIM environment, continuous capture, storage, and management of 3D digital data provides an efficient mechanism for monitoring and assessing every piece of information throughout the project lifecycle. Since BIM tools visualize the design model of a building in its early stages, a clear representation of the building is communicated to owners, which would subsequently facilitate modifications in line with requirements specified by the client.

3.2.1 BIM—Process and Product Perspectives

The widely used acronym BIM refers to two distinct, but closely linked, concepts, namely, product and process. As a process, BIM (Building Information Modeling) is an approach to creating, managing, deriving, and sharing building information among different AEC actors involved in various phases of the building development process to facilitate collaboration and communication among these actors (Eastman et al. 2011). BIM-based processes provide an integrated virtual environment encompassing all dimensions and disciplines associated with a built asset, which allows AEC actors to collaborate and communicate more efficiently and precisely compared to the traditional and ad hoc processes.

Consequently, the result of the BIM process is a BIM product (Building Information Model) which includes 3D digital spatial information, as well as semantic information about a building to support decision-making throughout its lifecycle (Smith and Tardif 2009). BIM products provide efficient spatial representation and communication of architectural, as well as structural, elements inside complex buildings (NBIMS 2012). There are multiple actors involved in the development of buildings, such as architects, civil engineers, MEP (Mechanical, Electrical, and Plumbing) engineers, and builders. Each actor can create and use their distinct BIM product. Throughout the design and construction phases of the BIM process, these distinct products can be assembled in a common data environment (CDE) which provides a complete model of the building (Bazjanac 2004). This complete model can subsequently be used by owners or facility managers for the purpose of operating and maintaining the building over the post-construction period. The CDE term refers to a centralized digital data repository in which an AEC project information is kept and maintained. The ultimate goal of a CDE is to enhance the generation, exchange, and dissemination of information that provides the basis for project delivery.

The main characteristics of a BIM product are (Isikdag et al. 2007; Isikdag 2015, 3):

- Object-oriented: Data stored in BIM models is based on object-oriented data structures. These data structures provide the capability to query and analyze BIM models for different applications.
- Information richness: BIM models include a rich data repository comprising both spatial and semantic information describing a wide range of building elements.
- Multi-dimensional: BIM models include not only a 3D visualization of building components but also manage other dimensions, such as time and cost, over the lifespan of a building (see Section 3.2.3).
- Spatially linked: A wide range of spatial relationships between building elements, such as connectivity, containment, and hierarchical aggregation, can be supported by BIM models.

- Extensibility: The open IFC standard and proprietary BIM authoring tools, such as Revit, provide extension mechanisms for enriching BIM models with additional data elements, which do not exist in the BIM model, in order to manage information required in any domain.
- Generation of model views: A subset of the BIM model can be extracted to meet the requirements of a specific group of AEC actors. These subsets are known as model views, and they can also be aggregated to form a combined model view, described in more detail in the following section.

3.2.2 BIM Maturity Levels

The purpose of maturity levels is to provide a simplified classification for adopting technologies and collaboration approaches to facilitate a better understanding of the processes, tools, and methods that can be used in an AEC project. As shown in Figure 3.1, the UK government has introduced four maturity levels (Level 0 to Level 3) for the BIM environment (Centre for Digital Built Britain 2011).

Level 0 is predicated on 2D CAD drawings which include points, lines, and polygons, as well as textual information. On this level, 2D digital or paper-based documents are used for exchanging and distributing building information. There are no common standards and processes at Level 0, which indicates collaboration among AEC actors is not efficient. For these reasons, the current AEC industry has now moved well beyond this level.

Level 1 comprises a structured CAD environment which includes a spatially coordinated and managed combination of 3D and 2D digital models. Examples include architectural design and duct fabrications models. Semi-structured digital documents are also used on this level to manage a wide

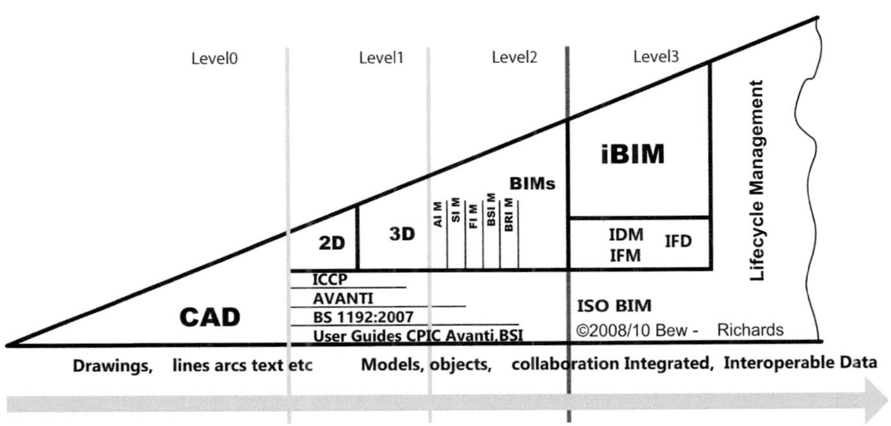

FIGURE 3.1
BIM maturity levels.

range of non-spatial datasets for the building components. Collaboration among AEC actors is predicated on file-based data exchange using a common data format. At Level 1, models may not be shared by all project stakeholders; thus, this level is sometimes called a "Lonely BIM." In other words, each discipline creates its own model at any stage of the project lifecycle (see Figure 3.1). Communication and exchange of information is in an ad hoc manner, resulting in disconnected and asynchronous models. Although institutional challenges and risk allocations exist at Level 1, detailed digital models provide the ability to accelerate the progress of an AEC project throughout its lifecycle. Therefore, by achieving Level 1 of BIM maturity, AEC actors will be encouraged to move towards model-based multi-disciplinary collaboration, as defined in Level 2.

Level 2 defines a federated or composite environment, in which BIM models created by various disciplines are logically harmonized and coordinated (see Figure 3.1). These BIM models include both 3D geometric and semantic information for a particular view of an AEC project. On Level 2, model-based collaborative processes should be considered to harmonize data flow among different systems and coordinate project stakeholders in the context of a specific project (British Standards Institution 2013). Model-based communication and data exchange can be done inside a particular phase or between two phases over the course of a project. For example, architectural and structural BIM models can be exchanged during the design phase of a project. Another example is exchanging data between an architectural model, which is created in the design phase, and a facility management model, which is used in the operation phase. At this level, there is a need to amend contracts, since document-based processes are replaced by augmenting model-based workflows. In addition, granularity of models is being increased as the project progresses forward.

At Level 3, models from various disciplines are connected and synchronized using an integrated network-based data-sharing space. (see Figure 3.1). Implementation of this level can be accomplished based on cloud platforms, distributed database management systems, and/or service-oriented architecture (SOA). Complex computations and analysis can be performed in the virtual phase of design. Real-world implementation of BIM level 3 has not yet been fully achieved in the AEC industry. The UK government provided the following key measures for realizing this level (Centre for Digital Built Britain 2015):

- Development of open data standards to facilitate data-sharing throughout the entire market.
- A new contractual framework should be established to prevent confusion, verify consistency, and facilitate collaborative and open procedures.
- Establishment of a cooperative environment with an effective sharing and learning culture.

- Educating non-expert clients in BIM-based techniques, such as data requirements, contractual procedures, and operational approaches.
- Driving domestic and international growth and jobs in technology and construction.

This level provides the instantaneous synchronization and exchange of models, which results in significant overlaps between various phases of the project lifecycle and eventually leads to phase-less processes. In other words, the fundamental assumption behind Level 3 is to enable "concurrent construction," which means all tasks and activities occurring in the project lifecycle are simultaneously planned and integrated to provide an optimal approach for designing, constructing, and managing a project (Succar 2009). Implementation of Level 3 requires substantial changes in the current way of thinking and collaboration in the AEC industry. There should be major changes in the current workflows, legal relationships among various parties, and so on.

3.2.3 Multi-Dimensional BIM (4D, 5D, and 6D)

The common perception of the BIM environment is that it is a collaborative process for creating and managing 3D physical dimensions of a built asset. However, non-physical dimensions of built structures can be also incorporated into the BIM environment. These extra dimensions differ from BIM maturity levels. They define a specific approach for linking specific types of datasets with a BIM model. Adding these additional dimensions would provide a more comprehensive understanding of an AEC project.

4D BIM is constructed based on intelligent linking of time or schedule-related information to a 3D BIM model (Popov et al. 2006). Therefore, 4D means 3D plus time (or schedule). 4D BIM models are capable of visualizing and displaying the entire progress of an AEC project over the course of its lifespan. The time dimension is added to physical elements over the progress of a project. It helps us to obtain an accurate visualization approach to communicating the sequence of activities that occur during the project lifecycle. Federating schedule-related information with 3D BIM models will help planners to develop an accurate program for an AEC project. Common examples of time-related information are the time needed for constructing a component, the required time for operating a component, and the sequence of installing a set of components.

5D BIM models add the additional dimension of cost to 4D models (Goedert and Meadati 2008). An accurate estimation of project costs can be determined using 5D BIM models. Cost-related information may include capital costs of buying and installing components, maintenance costs, and predicted replacement costs. Project costs can be calculated based on the associated information linked to building components. Cost managers can clearly estimate the cost quantities associated with a particular component, which would enable them to predict the overall cost of an AEC project. In 5D BIM models, predicted and real expenditure can be easily monitored and tracked over the project course. It should be noted that the accuracy of cost computations is predicated on the data generated by the project team. Inaccurate information will result in an erroneous estimation of costs.

6D BIM refers to incorporating additional information into 5D BIM models to enable the integrated management and maintenance of facilities during the building lifecycle (Pärn, Edwards and Sing 2017). 6D BIM is sometimes known as integrated BIM (iBIM) and covers all lifecycle aspects in facility management. This lifecycle information may include the constructor of a building component, maintenance details, comprehensive manuals about the optimal configuration and operation of each building component, energy performance, and so on. 6D BIM facilitates the pre-planning of activities required for maintenance and operation of an AEC project. 6D BIM models are typically handed over to the owner after completing the construction phase. Compared to the traditional folders, 6D BIM models provide a more efficient approach to monitoring, accessing, communicating, and updating information about built assets. Seamless exchange and flow of facility management information are supported in a 6D BIM environment, which would subsequently optimize the operational procedures in facility management practices.

3.3 Open BIM

Proprietary BIM environments would result in significant data interoperability issues during the project lifecycle. Consequently, open BIM environments have been developed to enable a universal data-sharing approach to the collaborative design, construction, and management based on open workflow procedures and standards. An open BIM environment enables various AEC actors to share and exchange their data irrespective of a proprietary BIM platform used internally within an organization. This environment provides a common language for a wide range of processes, providing a transparent approach to the procurement of AEC projects. The project dataset in an open BIM environment can be continuously used over the course of the project lifecycle, preventing the issue of creating the same datasets multiple times.

The BuildingSMART organization has been developing five basic standards to establish open BIM environments in the AEC industry (BuildingSMART 2017). These standards are: Information Delivery Manual (IDM), Model View Definition (MVD), International Framework for Dictionaries (IFD), BIM Collaboration Format (BCF), and Industry Foundation Classes (IFC). Among these standards, the IFC standard underpins the sharing of data across the project lifecycle. This standard is an open BIM data model which includes physical, spatial, and semantic entities describing the whole lifecycle of buildings. In the following subsections, all open BIM standards, except IFC, will be explained. A comprehensive description of the IFC standard will be presented in Section 3.4.

3.3.1 IDM

To unlock the full benefit of the BIM-based paradigm in the AEC industry, there should be efficient and enhanced communication among various project stakeholders in the building development processes. If the required information were accessible for use at the right time and its quality were satisfactory, there would be a significant improvement of various processes during the design, construction, and management phases of an AEC project. Although the IFC standard captures the totality of datasets to be exchanged within the project lifecycle, it does not provide adequate guidelines for capturing the subtleties of each individual business process.

The IDM standard provides an efficient methodology to enhance the quality of information communication by capturing and modeling processes and workflows during the lifecycle of an AEC project (Wix and Karlshoej 2010). Integration of *information* with *processes* would be considered as an anchor point in realizing the full potential value and benefits of the BIM environment for the AEC industry. IDM methodology comprises four major features:

- It specifies when a process fits for a particular purpose, justifying the relevance of the process.
- It assigns the roles of stakeholders who create, send, and receive the project data in each individual process.
- It defines the details of the information to be exchanged over the course of a process.
- It provides specifications for supporting the information exchanges by BIM tools.

Adopting an IDM-based approach would provide considerable benefits for both software developers and typical users of BIM. Software developers can use the IDM standard to break processes down into specific functional parts and specify the required capabilities of the IFC in support of each functional part. The IDM would enable developers to address user requirements more efficiently, assure the quality of the exchanged data, and reuse software components multiple times. Since the IDM adopts a plain language to describe processes, every BIM user can obtain a clear understanding of business processes occurring in the various phases of design, construction, and operation of buildings. This would enable more reliable mechanisms for exchanging information among project actors, facilitating decision-making in complex situations.

There are three types of user targeted in implementing the IDM standard: an executive user, an end user, and a solution provider (Wix and Karlshoej 2010).

An executive user decides whether to adopt an IFC-based approach for exchanging BIM data. Although these users are not concerned about the details of information exchanges within each process, they have a business perception of these processes and are conscious of process enhancements and potential impacts on the business level. An executive user typically does not require the technical knowledge of application development or data structure of the IFC standard. End users are involved in the practical usage of IFC files. Therefore, they should have a good understanding of the details of datasets exchanged and used in each business process. There is no need for end users to develop software tools or understand the data model behind the IFC standard. Finally, solution providers are people or companies involved in the development of IFC-based applications or interfaces. These providers should have a good understanding of the expectations of both the executive and end users from their software tools. Therefore, they must have proficient skills in the technical implementation of the IFC standard.

The IDM comprises four main components: process maps, exchange requirements, functional parts, and concepts. A process map comprises the workflow of tasks and activities conducted for a particular purpose. It aims to provide a clear image of the arrangements of activities and tasks in terms of the involved actors and their roles, as well as the required information exchanges between these actors. An exchange requirement is defined as a set of datasets required to be exchanged to meet a specific requirement during any stage of the project lifecycle. These datasets should typically be created in line with the data structure of the IFC standard. However, it is also possible to adopt an IDM-based approach for other open data standards, such as Geography Mark-up Language (GML), which lays the foundation for open data exchanges in the geospatial domain. A non-technical language is normally used for describing exchange requirements. BIM users, such as architects, engineers, and builders, are the major audience for whom an exchange requirement is expressed. However, solution providers should also utilize exchange requirements, since they provide the basis for developing software tools in alignment with user needs.

A functional part is a consistent set of reusable information units that solution providers should use to meet a particular exchange requirement. Functional parts are composed of concepts. In other words, a concept is a piece of information embedded in a functional part. It is not essential to associate a concept with a whole IFC entity. It can also describe the approach to modeling attributes of an IFC entity, such as naming conventions, global identification, and so on.

3.3.2 MVD

A Model View Definition (MVD) is defined as a subset of the IFC schema, which supports one or several exchange requirements defined in an IDM. The main role of an MVD is to ensure that an implementation of the IFC standard meets the end user's needs, as specified in the form of exchange requirements within an IDM document. In addition, the main aim of an MVD is to verify that the specified subset of the IFC standard is completely implementable in both sending and receiving BIM platforms. There are three main deliverables that need to be considered in each MVD (See and Karlshoej 2012):

- Overview or description of the MVD: This defines the scope of the MVD in terms of the version of, and the IFC independent, MVD, basic principles in mapping the MVD onto a particular release of the IFC standard, and limitations considered in the MVD definition.
- MVD diagrams: These diagrams provide a schematic representation of the data structure of the concepts, as well as their logical relationships as specified in the exchange requirements.
- Guidelines and specifications for implementing concepts: These guidelines explain how entities from the IFC standard can be used to implement each concept in an IFCbased information exchange.

There are some requirements which should be met by any MVD. First, data exchanges, which are defined in the IDM processes, must not be ambiguous. The MVD may modify and change some data exchange requirements into meaningful packages in terms of software implementation. Another requirement is that the MVD should provide an optimal approach for software developers to implement IFC-based information exchanges among BIM platforms. There must be no need for conducting face-to-face meetings or workshops for development and implementation of IFC support in pre-defined data exchange requirements. An MVD-based certification must provide adequate information about the existing capabilities and restrictions of the implemented IFC-based information exchanges. Professionals in the AEC industry should be sure to understand the content of IFC data exchanged via an MVD. In other words, what sort of IFC data can, or cannot, be exchanged between receiving and sending BIM platforms.

MVDs provide the users with the ability to exchange open BIM data in business processes associated with real-world projects. Therefore, it must be ensured that exchange requirements defined from an end user perspective are implementable in the software environment. An important deliverable of any MVD implementation should be a single page describing the content of the MVD. All team members involved in implementing MVDs should review this description page for already developed MVDs in order to determine if there is a need for developing a new MVD or if the existing one can be extended in line with the exchange requirements.

MVDs will come into existence if software developers implement them in BIM platforms. Therefore, the MVD development team should make sure that software vendors tend to implement the MVD. A key element of concept definitions in MVDs is implementation agreements, which provides more specific requirements for implementing concepts. This is helpful when these concepts are defined vaguely or broadly in the overall schema of an IFC standard. For instance, there might be six geometric representations considered for a particular concept in an IFC standard; however, an implementation agreement would restrict this to one or two geometric representations which are adequate in the context of an MVD.

A large number of teams should be involved in developing IDMs and MVDs to encourage the wide adoption of the IFC-based BIM data exchange in AEC projects around the world. Therefore, coordination among these teams is important, which would enable software developers to find and reuse existing concepts and register new concepts for use by others. MVDXML is a format used for encoding concepts defined within MVDs. The aim of the MVDXML format is to automate the process for validating and checking IFC data, which would subsequently eliminate the need for error-prone manual checking processes. The BuildingSMART organization has developed and published the official release of a number of MVDs for IFC2x3 and IFC4 versions of the IFC standard. The released MVDs for IFC2x3 are IFC2x3 Coordination View Version 2.0, IFC2x3 Coordination View, IFC2x3 Structural Analysis View, and IFC2x3 Basic FM Handover View. For IFC4, two MVDs have been published: IFC4 Reference View and IFC4 Design Transfer View.

3.3.3 IFD

To automate the verification process for information exchanges, the general level of the IFC standard should be detailed with more information. For instance, an architect may use a plain text string to provide information about the materials inside walls. Even though this information is spelt correctly, there is no assurance the receiving BIM platform can interpret it correctly. What would happen if information in the sending application is encoded in a language which is different to the one used in the receiving application? These issues refer to the semantic interoperability in the BIM environment. Therefore, the semantics behind the exchanged information needs to be defined.

The BuildingSMART organization has developed the IFD standard to resolve the interoperability issues at the semantic level. The IFD standard provides an open, shared, and international framework for defining the terminology or vocabulary used in open BIM-based information exchanges (Bell and Bjørkhaug 2006). This standard provides semantic descriptions of all concepts and terms used in open BIM environments.

More recently, IFD is known as the BuildingSMART Data Dictionary (bSDD). The core part of the bSDD technology is the Global Unique Identifier (GUID). GUIDs serve as unique, language-dependent serial numbers—each assigned to every concept you may find in bSDD. It is to these GUIDs that the various language terms and definitions are associated. While human actors involved in an information exchange can read and interpret strings for names and descriptions, computers can communicate and interpret the meaning of information correctly by relying on GUIDs. The GUID is an identifier that is unique throughout the software world. GUIDs are created by using an algorithm which virtually guarantees the uniqueness of each GUID. This GUID enables bSDD users to be "assured" that when a concept is used, a different concept from any other source will not share the same identity.

While the IFC standard provides the entities for modeling objects, defines the connection between these objects, and describes the storage of building information, the bSDD standard provides a unique description of objects and their constituting parts, properties, units, and values.

There are two types of content defined within the bSDD: concepts and characteristics. A concept is simply any "idea" which is normally expressed through a language-dependent name. Concepts may describe an object—real or abstract, a property about an object, a relationship, and so on. There might be several labels for one independent concept in various languages or even the same language. Conversely, a name might be used as a label for several concepts. Names and independent concepts have their own GUIDs that are connected via a naming relationship. Characteristics are a specific type of concepts that cannot be defined using other concepts. A description is used for defining their meaning. There are various types of characteristics including property, unit, behavior, function, and measure.

Using the bSDD standard provides significant benefits including:

- Smoothing the understanding and flow of information in the open BIM environment.
- Reduction in costs and improvement of quality due to the reusable object libraries shared among project stakeholders.
- Any software tool can have access to the same definition of objects and properties.
- The bSDD standard provides the basis for automatic rule-checking and validation, which subsequently prevents miscommunication and data duplication.
- Objects can be connected to BIM-specific products.
- Property sets defined in the IFC standard can be extended to meet specific project requirements.

3.3.4 BCF

Prior to 2010, BIM end users had to share the whole BIM model with each other to communicate issues, request changes, and propose suggestions inside open BIM environments. As a result, different versions of the bulk BIM model were required to be reviewed by the receiving member to filter the change request from the sending person. This has resulted in significant communication delays over the workflow of AEC projects. To mitigate this challenge, a proposal for developing an open standard has been initiated to coordinate changes and communications among various BIM tools.

BIM Collaboration Format (BCF) is an open XML-based data format used for keeping change coordination and communication information apart from open BIM models (Stangeland 2013). In other words, the IFC standard enables the structure and exchange of 3D lifecycle information, while the BCF provides an efficient mechanism for managing communications around open BIM models. BCF files typically provide details about each task, which is called a *topic* in the context of BCF. A BCF-based topic provides task status, assignee and task type, all comments associated with the task, and references to relevant components inside the BIM model. BCF provides a simplified approach for encoding and coordinating change messages, which would subsequently facilitate communication among project members working on an open BIM model. BCF provides the project team with the ability to raise, respond to, and comment on issues within an open file format that does not itself contain components of the exchanged BIM model. Therefore, the major role of BCF is to separate the "communication" from the "model" in open BIM environments. The implication of using BCF is to promote the level of collaboration, since this format provides the ability to exchange very small amounts of data, rather than the whole BIM model, among the team members of a project.

The first version of BCF, known as "bcfXML v1," was initiated by the Solibri and Tekla companies in 2010. This was then followed by the formation of a task force under the Implementer Support Group of BuildingSMART to develop a new version called "bcfXML v2." This version provided improved capabilities, such as exchanging machine readable BCF information, attaching small parts, viewpoints, or snapshots of a BIM model.

In real-world projects, it is essential to exchange a very large number of BCF files to finalize all tasks over the building development process. In this case, managing BCF files could be challenging for the project team. In response to this challenge, BCF web services were introduced to communicating BCF-based topics using the cloud through a RESTful API (Application Programming Interface) known as bcfAPI, which was developed for the BCF standard. bcfAPI enables an automatic synchronization of updates in BCF-based topics shared within the project team.

3.4 IFC Standard

IFC provides a comprehensive data structure for storing BIM data in a digital repository which can be shared and exchanged between IFC-compliant BIM tools. In other words, this standard is the underlying open schema for storing and managing building information within the collaborative environment of BIM (ISO16739 2013). It is considered an open data specification which

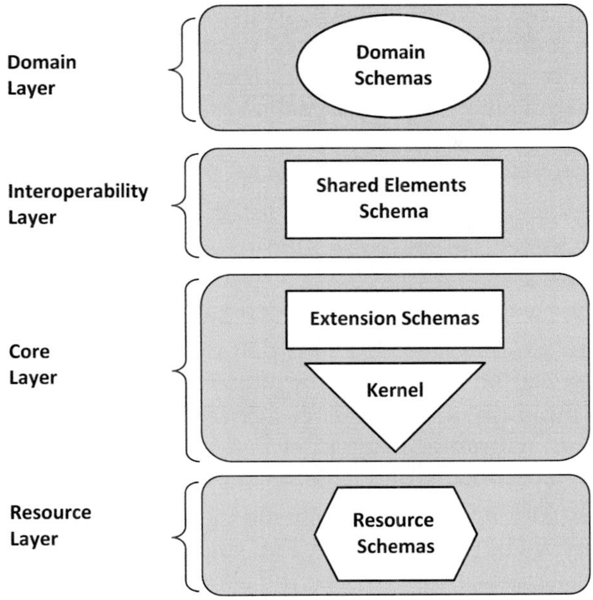

FIGURE 3.2
Modular structure of the IFC standard.

is independent from a particular BIM vendor. As indicated in Figure 3.2, the schema of the IFC standard is modular and includes four main conceptual layers, (ISO16739 2013), namely, a resource layer, a core layer, an interoperability layer, and a domain layer. Each IFC layer comprises several subschemas. In the following subsections, a comprehensive description of each layer will be presented.

3.4.1 Resource Layer of IFC

The resource layer is the lowest layer in the conceptual data schema of the IFC standard. This layer includes subschemas which hold basic concepts and generic entities (ISO16739 2013). Resource level entities cannot exist independently, and their existence is dependent on using them as value type of attributes in higher level entities which are derived from an "IfcRoot" class. The subschemas relevant to the scope of this book are explained in the following (BuildingSMART 2013f):

- Actor Resource: This subschema provides the capability to model information describing the actors, including people and organizations, which are associated with various tasks and activities (ISO10303-41 2005). Additionally, relationships between individuals and their organizations and relationships between two organizations are also defined within this subschema to support hierarchical institutional structures.

- Approval Resource: This subschema includes the fundamental concepts required in the approval processes, such as for a plan, a design, a proposal, or a change order in the construction or facilities management project. Information regarding requesting and giving approvals, as well as approval conditions, can be defined by IFC entities provided in this subschema.
- External Reference Resource: This subschema enables IFC entities to be linked to external sources of information such as documents, external data libraries, and classification systems. Also, metadata attributes about these external resources are defined within the IFC schema.
- Geometric Resource Model: This defines various types of geometric representation approaches for modeling the geometric or spatial form of building elements. These geometric models include CSG, half-spaces, manifold B-rep, faceted B-rep, swept solids, bounding boxes, surface models, and tessellated models.
- Geometry Resource: This subschema includes entities for modeling various types of points, curves, and surfaces which can be used to define the spatial extent of building elements. Moreover, these entities can be used to describe geometric models defined within the "Geometric Resource Model" subschema.
- Measure Resource: Various types of measurement units and types, which are typically assigned to quantities, are defined within the Measure Resource subschema. Examples are SI units such as meter, square meter, and cubic meters, used for measuring length, area, and volume of spatial objects, respectively.
- Topology Resource: This subschema comprises entities used for defining and orienting topological primitives such as vertex, edge, and face. Additionally, a specialized subtype is defined for each vertex, edge, or face to provide their association with the geometry of a point, curve, or surface, respectively. Basic topological entities can be aggregated to form spatial topological structures of the path, loop, and shell and the integrity of these structures can be validated using constraints.

3.4.2 Core Layer of IFC

The core layer consists of the "IfcKernel" subschema and three core extension subschemas, namely product, process, and control extension. "IfcKernel" contains the most abstract entity "IfcRoot" which is specialized into three fundamental and abstract entities (BuildingSMART 2013g):

- Object definitions (IfcObjectDefinition): This entity is the abstract superclass for entities, capturing different types of objects, such as products (IfcProduct), processes (IfcProcess), and controls (IfcControl).

- Relationships (IfcRelationship): This entity is the abstract superclass for defining all relationships among objects in IFC. There are six types of relationships: composition, assignment, connectivity, association, declaration, and definition. These relationships are also further specialized in order to define more specific relationships. For instance, "IfcRelSpaceBoundary," which is a subclass of the "IfcRelConnects" entity defining the general concept of connectivity relationships, defines connectivity relationships more specifically between space objects and building elements.
- Property definitions (IfcPropertyDefinition): This concept defines the generalization of specific groups of properties. Here, property refers to a user-defined attribute that can be assigned to objects via relationships. The subclasses of "IfcPropertyDefinition" are occurrences of property sets, property templates, and property set templates. These entities will be useful for extending and specializing general IFC entities without the need to define a new subclass. For instance, the "IfcActor" entity can be extended with property set templates to model information related to land surveyors. The naming convention for a group of property sets is "Pset_xxx."

Product extension of the core layer defines the fundamental concepts for modeling products (BuildingSMART 2013c). Here, the term "product" is the generalized concept for those objects which likely have a geometric representation and spatial placement. These objects mainly include various spatial structures, physical elements, and space elements (see Section 3.4.8). In addition, the grid (IfcGrid), which imposes constraints for the placement of elements, and annotations (IfcAnnotation), such as explanatory text and dimensioning, are defined as specializations of the "IfcProduct" entity.

Process extension of the core layer builds the foundation for modeling concepts required to map processes in terms of logical planning and scheduling of required activities and tasks (BuildingSMART 2013b). These concepts include a definition of work plans (IfcWorkPlan), schedules (IfcWorkSchedule), events (IfcEvent), procedures (IfcProdecure), and tasks (IfcTask). Control extension of the core layer comprises basic entities required for controlling objects including products, processes, or resources (BuildingSMART 2013a). More specific control entities include performance histories (IfcPerformanceHistory) and approval associations (IfcRelAssociatesApproval). "IfcPerformanceHistory" is adopted to document the performance of building elements over a specified period. It provides machinegenerated data resulting from building automation systems or human-specified data resulting from manual tasks or activities. "IfcRelAssociatesApproval" is utilized to relate information for approvals to any object, such as documents.

3.4.3 Interoperability Layer of IFC

The interoperability layer specifies subschemas which contain those IFC entities shared, used, and exchanged across multiple AEC domains. Current shared subschemas include (BuildingSMART 2013i):

- Shared building elements (IfcSharedBldgElements): This subschema defines the specialized entities of the abstract "IfcBuildingElement" entity such as walls (IfcWall), doors (IfcDoor), windows (IfcWindow), slabs (IfcSlab), roofs (IfcRoof), and similar elements. These elements form the basic architectural configuration of buildings.
- Shared building service elements (IfcSharedBldgServiceElements): This consists of various entities used for modeling distribution elements required in building services such as electricity, water, and heating. Examples of distribution elements are distribution chambers (IfcDistributionChamberElement), flow controllers (IfcFlowController), flow fittings (IfcFlowFitting), flow segments (IfcFlowSegment), flow terminals (IfcFlowTerminal), and so on.
- Shared component elements (IfcSharedComponentElements): Varying small components of buildings, such as accessories and fasteners, are modeled within this subschema.
- Shared facilities elements (IfcSharedFacilitiesElements): Basic concepts of facility management are modeled in this subschema. These concepts can be used by applications which require sharing information about facilities management-related issues.
- Shared management elements (IfcSharedMgmtElements): This subschema comprises the major entities needed for the management of projects throughout the various stages of the building lifecycle. These entities provide the capability to capture and control information about scope, cost, and time of projects.

3.4.4 Domain Layer of IFC

The most specific subschemas for each AEC domain are defined within this layer. Entities defined in these subschemas are self-contained, and there is no possibility to further specialize or reference them in other layers (ISO16739 2013). Currently, the domain layer includes subschemas for eight AEC disciplines, namely, the building controls domain (IfcBuildingControlsDomain), the plumbing and fire protection domain (IfcPlumbingFireProtection Domain), the structural elements domain (IfcStructuralElementsDomain), the structural analysis domain (IfcStructuralAnalysisDomain), the heating, ventilation, and air conditioning (HVAC) domain (IfcHvac Domain), the electrical domain (IfcElectricalDomain), the architectural domain (IfcArchitectureDomain), and the construction management domain (IfcConstructionMgmtDomain).

3.4.5 Object Placements and Spatial Reference Systems within the IFC Standard

As the most abstract entity for representing spatial objects in the IFC schema, "IfcProduct" references the "IfcObjectPlacement" entity through its "ObjectPlacement" attribute to define the spatial placement of product elements, including spatial and physical elements (Liebich 2009). The abstract "IfcObjectPlacement" entity is specialized into two instantiable subclasses (BuildingSMART 2013e):

- IfcLocalPlacement: This entity provides two ways to define the spatial placement of products. The first approach is to set an absolute placement of a product with respect to the world spatial reference systems, such as WGS (World Geodetic System) 84. Another option is to set a relative placement of a product on the local spatial reference system of another product. For example, levels of a building (IfcBuildingStorey) can be spatially placed with reference to the whole building object (IfcBuilding).
- IfcGridPlacement: This entity defines the spatial placement of a product by defining a constraint placement relative to the axes of a design grid (which is defined by the IfcGrid entity). In fact, the location of a product is defined by using a virtual intersection and reference directions provided by axes of the grid.

Since the IFC standard includes various options to define spatial reference systems both locally and globally, BuildingSMART suggests the following recommendations for defining spatial placement of main product elements (BuildingSMART 2013d):

- IfcSite: Sites shall be placed with respect to the world coordinate system.
- IfcBuilding: Spatial placement of buildings shall be defined relative to the local placement of the site on which they are built.
- IfcBuildingStorey: Building stories shall be placed with respect to the local placement of their corresponding building.
- IfcElement: Spatial placement of physical elements can be defined in various ways, depending on their spatial relationships with other spatial or physical elements. One case is defining containment relationships between physical elements and either sites, buildings, or building stories. For example, geographic elements, which are defined by IfcGeographicElement (a subclass of IfcElement), can be spatially contained with sites (IfcSite). In this case, a geographic element should be spatially referenced with respect to the local placement of its container (site). In another example, walls (IfcWall is a

subclass of IfcElement) may be spatially contained within building stories. Similarly, spatial placement of walls is defined with respect to their container (building story).

The above recommendations indicate that "IfcSite" is the central entity used for defining geospatial reference systems and placement of other elements is defined relatively within the IFC. There are three attributes (RefLatitude, RefLongitude, RefElevation) in the "IfcSite" entity. These attributes define latitude, longitude, and datum elevation from sea level for the reference or origin point of each site, respectively (BuildingSMART 2013j). The origin point of the site is (0, 0, 0) within its local reference system. However, its global placement is typically predefined based on geodetic coordinate values within WGS84. To define WGS84 and other world spatial reference systems, two main concepts were considered in the IFC schema (Liebich 2009):

- The concept of spatial reference systems: There are two entities for modeling this concept: the "IfcCoordinateReferenceSystem" and its instantiable subclass "IfcProjectedCRS." "IfcCoordinateReferenceSystem" entity models both the geodetic datum and vertical datum used in defining a world spatial reference system. The geodetic datum indicates the shape and size of the rotation ellipsoid and its connection and orientation to the real shape of the Earth. This datum is used for defining geographic coordinates, namely latitude and longitude. Examples of a geodetic datum are WGS84, North American Datum 1983 (NAD83), European Terrestrial Reference System 1989 (EUREF89), and Hayford. A vertical datum indicates the reference plane and base point defining the origin of a height system. Examples of a vertical datum are the Mean Sea Level (MSL), and North American Vertical Datum of 1988 (NAVD88). "IfcProjectedCRS" is used for defining projected spatial reference systems. Projected spatial reference systems, which are defined by using map projections, convert geographic coordinates to map coordinates (X, Y). The "IfcProjectedCRS" entity includes three specific attributes, namely *MapProjection*, *MapZone*, and *MapUnit*. The *MapProjection* attribute provides the name of the map projection, such UTM (Universal Traverse Mercator) or Gaus-Krueger. *MapZone* specifies the zone number of the map project. For example, 38 can be assigned for defining UTM38. *MapUnit* defines length units (e.g. meter) assigned to the coordinate axes of the projected spatial reference system.
- The concept of coordinate transformation operators: Two IFC entities mainly define transformation operators, including the abstract "IfcCoordinateOperation" and its instantiable subclass "IfcMapConversion." These entities can also be used to transform coordinates from the local spatial reference system of the project into the world spatial reference system. "IfcCoordinateOperation" is the

abstract entity for defining transformation or conversion between any two spatial reference systems. "IfcMapConversion" more specifically defines the transformation from the local origin of the local spatial reference system to its place within a map (easting, northing, orthogonal height) and rotates the x-axis of the local spatial reference system within the horizontal (easting/westing) plane of the map.

3.4.6 Solid Models within the IFC Standard

For modeling the shape of spatial elements and physical building elements, "IfcProduct" references the "IfcProductRepresentation" entity through its *Representation* attribute (Liebich 2009, 121). This entity could refer to those entities defined in the "Geometric Resource Model" subschema to use solid models for defining the shape of 3D objects. Currently, IFC supports various types of solid models (IfcSolidModel). The most widely used solid models include B-rep (IfcManifoldSolidBrep), CSG (IfcCsgSolid), and Swept Solid (IfcSweptAreaSolid). In the next subsections, these solid modeling approaches are briefly described.

3.4.6.1 Constructive Solid Geometry (CSG)

The CSG solid modeling approach defines the geometric shape of 3D spatial objects by applying Boolean operators to the standard primitive objects (Rossignac and Requicha 1999). These primitive objects usually include simple shapes such as cuboids, cylinders, pyramids, spheres, cones, and so on. The main Boolean operators used in CSG are union (∪), intersection (∩), difference (–), and geometric transformations such as translation, rotation, and scaling. CSG solid models can be expressed as ordered binary trees, where the primitive objects are represented as leaf nodes. Results of Boolean operators are defined within the internal nodes, and the final CSG model is represented as the root node of the tree. Figure 3.3a shows an example of CSG solid model defined by applying Boolean operators on primitive objects.

3.4.6.2 Boundary Representation (B-rep)

B-rep solid models are merely defined based on a set of connected boundary surfaces (see Figure 3.3b). These boundary surfaces or faces are formed by a graph of edges and vertices (Allen 1984). At least one shell is required to show the final representation of a B-rep solid model. One shell, the outer, shall completely contain all the other shells and no other shell may include a shell. The Eular-Poincaré formula defines the following quantitative relationship among the number of faces (F), edges (E), vertices (V), faces' inner loops (L), shells (S), and genus of shells (G) in B-rep solid models (Mantyla and Sulonen 1982):

$$F - E + V - L = 2(S - G) \tag{3.1}$$

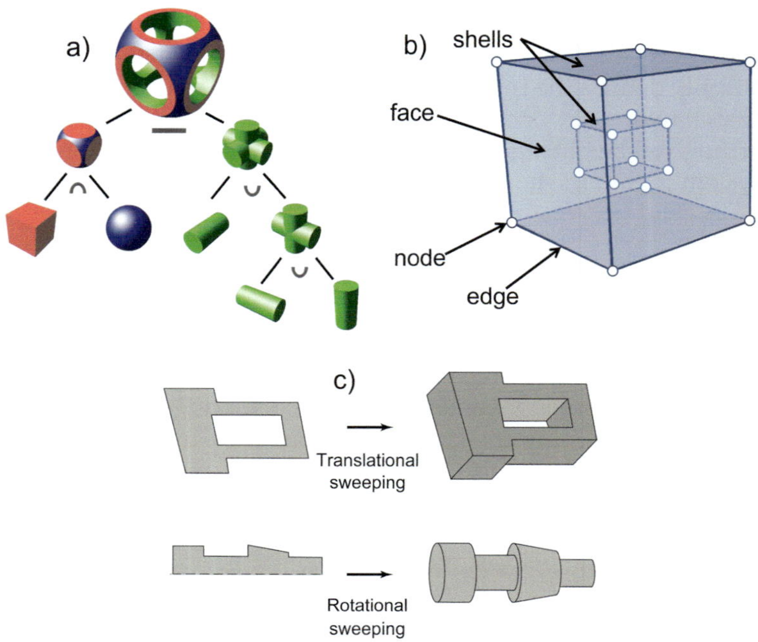

FIGURE 3.3
Widely used solid models (a) Example of CSG solid model (created by Wikipedia 2014); (b) Example of B-rep solid model (adapted from Nagel 2014, 171); (c) Examples of Swept solid models.

3.4.6.3 Swept Solid

The fundamental idea behind swept solid models is to represent them by a 2D profile and a predefined curve (Agoston 2005). The 2D profile could be a primitive object such as a rectangle, circle, or polygon. The volumetric extent of swept solid models is formed by either rotating or translating the 2D profile alongside the trajectory of the predefined curve (see Figure 3.3c). Some building elements, such as walls, columns, or beams, can be defined by applying translational swept solid models.

3.4.7 IFC Standard and Other 3D Spatial Data Models

Besides IFC, there are other 3D data models developed for exchanging 3D spatial information. These 3D data models can be classified into three main categories: purely geometric models, models with structured geometry but limited semantic information, and models comprising structured geometry as well as comprehensive semantics (Kolbe, Nagel, and Stadler 2009; Stadler and Kolbe 2007). VRML (ISO 2004), X3D (ISO 2005), COLLADA (ISO/PAS17506 2012), and KML (Wilson et al. 2007; Wilson 2008) are examples of the first category. The geometry of models in the second category can be constructed through automatic interpretation of photogrammetric data or laser scan point clouds (Kolbe, Nagel, and Stadler 2009). The constructed geometry can then be enriched with limited semantic information. Examples of the last category are CityGML as well as IFC (Kolbe 2012). IFC and CityGML standards are not only standards for exchanging 3D data, but also define the spatial data models for BIM and 3D GIS domains, respectively.

Although both IFC and CityGML can be used for modeling the spatial and semantic elements of buildings, they differ from each other significantly in terms of how they semantically model the built environment (Gröger and Plümer 2012). The focus of IFC is on the design and construction aspects of buildings such as walls, ceilings, beams, and slabs. IFC models typically use a parametric modeling approach to construct 3D spatial objects. Therefore, the whole building element is semantically assigned to one entity (such as IfcWall for a wall element). In contrast, CityGML defines an observational view of building elements and building features are modeled in terms of their observable surfaces (Nagel, Stadler and Kolbe 2009). Therefore, different semantic entities are assigned for observable surfaces of a building element. For example, "WallSurface" and "InteriorWallSurface" are respectively used for semantic modeling of external and internal surfaces of a wall element (see Figure 3.4).

Here, a comparison between IFC and its opponent (i.e. CityGML) will be presented. First, CityGML only utilizes the B-Rep method for representing the geometry of 3D objects (Kolbe 2012). However, IFC is more flexible in modeling the geometry of 3D objects, and it supports not only B-rep, but also other types of solid models, such as CSG and swept solids (BuildingSMART 2013h). B-rep more effectively supports 3D topology relationships in comparison with CSG (Ekberg 2007); however, CSG is more flexible in constructing complex spatial objects (Jarroush and Even-Tzur 2004). For instance, CSG can efficiently model a spatial object with a hole inside in comparison with B-rep. To illustrate this

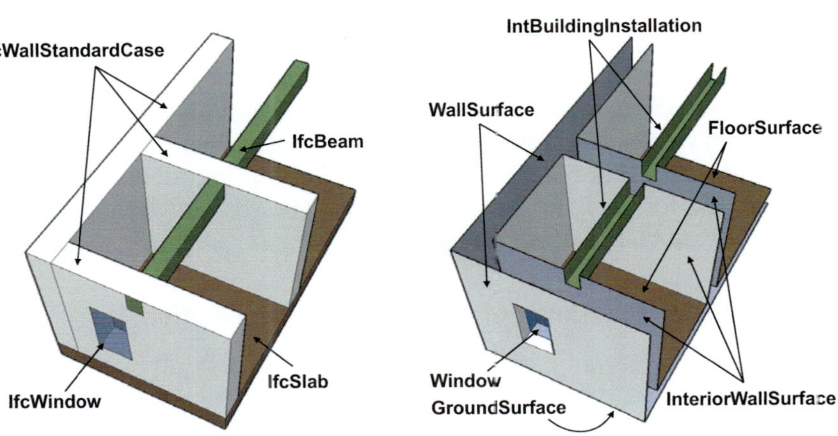

FIGURE 3.4
Semantic definition of building elements in IFC (left) and CityGML (right). (Adapted from Nagel, Stadler, and Kolbe 2009.)

FIGURE 3.5
A high-rise building with 3D holes (the Elephant building located in Bangkok, Thailand).

situation, a real-world building with two holes has been shown in Figure 3.5. These holes can be created in CSG by using one subtraction operation for each hole, which is subtracting the whole building object from the shape of the hole. In contrast, if the building model is constructed using the B-rep model, topological primitives must be appropriately defined—namely vertices, edges, and faces—which is a time-consuming and difficult task.

The second reason is that although LoD4 of CityGML provides all the entities needed to represent legal information, the data of LoD4 does not exist for most cities. Another reason is that in the design phase of a building subdivision process, land surveyors currently use 2D CAD files, which they receive from architects, as the basis for creating subdivision plans for high-rise buildings (Ho, Rajabifard, and Kalantari 2015). It is highly likely that this workflow would continue, i.e. if land surveyors are to create a 3D digital plan of subdivision for a building, they would likely request a 3D digital model of the building from architects. Currently, given that 3D building data in an IFC format is very prevalent among architects (Shojaei 2014), it is perhaps more feasible that land surveyors would receive 3D building models from architects in an IFC format in future, rather than CityGML, which they might use to create 3D digital building subdivisions.

3.4.8 IFC Entities Pertinent to Urban Land Administration

In this part, the relevant parts of the IFC schema will be shown using notations of EXPRESSG diagrams. To understand these notations, the reader is referred to Appendix A. The IFC entities relevant to the context of urban land administration can be classified into four main categories (Atazadeh et al. 2017; Liebich 2009): spatial elements, physical elements, document referencing elements, and actor assignment elements.

3.4.8.1 Spatial Elements

The "IfcSpatialElement" entity is the superclass for spatial elements. These elements are used for modeling various spatial structures within an IFC project (see Figure 3.6). Spatial structures can be defined in either a non-hierarchical or a hierarchical way. Spatial zones (IfcSpatialZone) are non-hierarchical decompositions of an IFC project. For hierarchical spatial structures, there are two main abstract entities:

- "IfcSpatialStructureElement" is the abstract superclass for entities defining site (IfcSite), building (IfcBuilding), building story (IfcBuildingStorey), and internal space (IfcSpace) objects.
- "IfcExternalSpatialStructureElement" is the abstract superclass for the "IfcExternalSpatialElement" entity which is used for modeling external regions and airspace around the building site.

3.4.8.2 Physical Elements

"IfcElement" is the abstract superclass for entities modeling physically existent objects (see Figure 3.6). The relevant physical objects are building elements (IfcBuildingElement), distribution elements (IfcDistributionElement), and geographic and civil elements (IfcGeographicElement and IfcCivilElement). Building elements constitute the primary parts of the buildings. These elements include windows (IfcWindow), doors (IfcDoor), walls (IfcWall), slabs (IfcSlab), stairs (IfcStair), and similar components. Distribution elements represent different types of utility networks inside the building, as well as around the building site. Examples are pipes, cables, and ducts. Geographic and civil elements represent geospatial features such as roads, trees, bridges, or pavements. Figure 3.7 represents examples of some spatial and physical elements.

3.4.8.3 Document Referencing Elements

Documents can be related to objects via the "IfcRelAssociatesDocument" relationship. As external resources, documents can be referenced and metadata for these documents can be managed within IFC. The resource layer of IFC includes the "IfcExternalReferenceResource" subschema which links IFC files to external resources. Within this subschema, two main entities provide the basis for managing documents:

- "IfcDocumentInformation" specifies the metadata about documents, such as a reference URL for the original file of the document.
- "IfcDocumentInformationRelationship" is an objectified relationship that relates documents to each other.

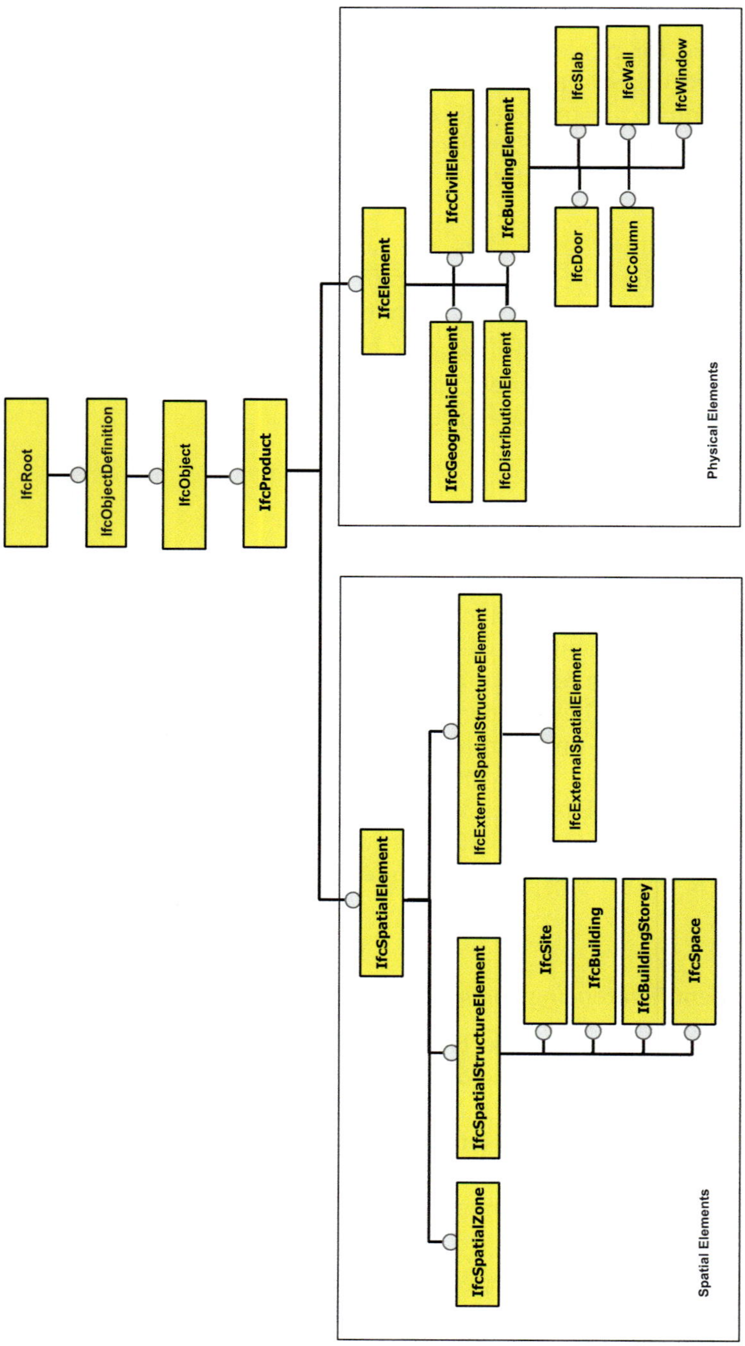

FIGURE 3.6
Relevant spatial and physical elements in IFC schema.

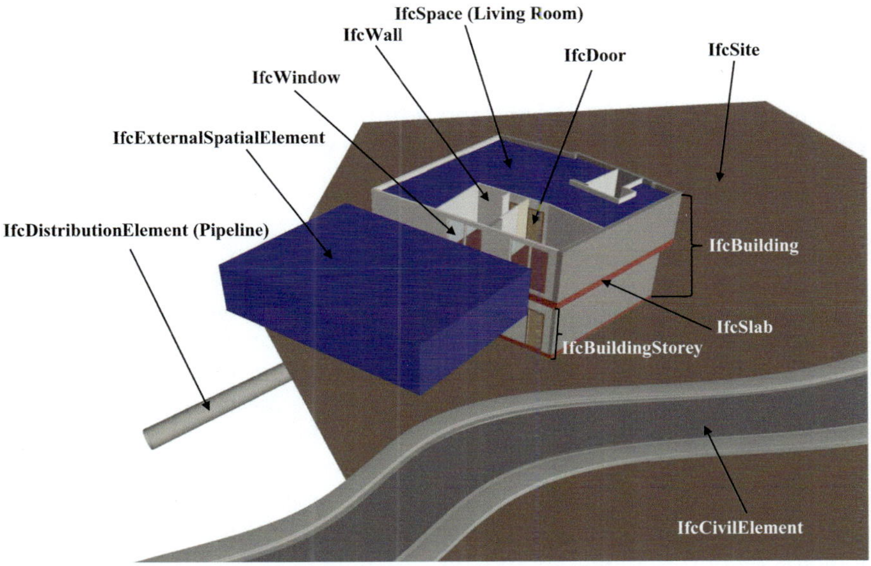

FIGURE 3.7
An indicative example that illustrates relevant spatial and physical IFC entities.

3.4.8.4 Actor Assignment Elements

In the core layer of IFC, the "IfcActor" entity provides the basis for defining various actors involved in the development process of a building project. Three types of actors can be selected via the "IfcActorSelect" attribute: a person (IfcPerson), an organization (IfcOrganization), or a person associated with an organization (IfcPersonAndOrganization). These entities are defined in the "IfcActorResource" schema within the resource layer. The role performed by actors is defined through the "IfcActorRole" entity.

3.5 BIM Projects in Land Administration

The benefits of using BIM models in the geospatial domain have been demonstrated for different urban applications, which require both indoor and outdoor spatial information, such as evacuation (El-Mekawy, Östman, and Hijazi 2012) and fire response management (Isikdag, Underwood, and Aouad 2008). One of the opportunities of bringing BIM into the geospatial domain is to enable the 3D registration of ownership rights in apartments in a digital environment and to facilitate land management in urban areas (Atazadeh, Kalantari, and Rajabifard 2016). This implies that land administration, as another area of the geospatial domain, can also utilize BIM for meaningful communication of complex 3D RRR spaces defined inside multi-story buildings (Frédéricque, Raymond, and Van Prooijen 2011; Isikdag, Zlatanova, and Underwood 2011). However, there remains limited research and practice into the feasibility of BIM for urban land administration. This section reviews existing preliminary projects to highlight the use of BIM for land administration in various jurisdictions around the world.

3.5.1 IFC for an Indoor Cadaster

One of the earliest projects on utilizing BIM for land administration was conducted by Clemen and Gründig (2006), who indicated the need to import different entity types into the IFC schema based on processed surveying measurements and observations to manage indoor cadastral information. In this investigation, it was demonstrated that as the IFC data model can support 3D topology and geometric representation of building elements, it could potentially be extended for land surveying purposes by pre-processing raw surveying measurements and observations. The authors emphasized that topology of observations should be distinguished from object topology already defined in the IFC schema. As a result, they proposed a list of those concepts which needed to be considered in the indoor cadastral domain of the IFC standard. These concepts include:

- Measurement units and conventions: Examples are clockwise or counter-clockwise measurements of angles, angles expressed in radiant grad or degree, and right- or left-handed spatial reference systems.
- Raw observation value: Examples of this concept are direction, distance, zenith angle, target height, uncertainty.
- Observation node: This concept refers to geodetic or surveying points.
- Observation edge: This concept connects two observation nodes with each other and is associated with a set of raw observation values.
- Observation group: This concept can model a set of observation edges with the same translation, orientation, and time stamp.

To manage cadastral information in indoor environments, the aforementioned concepts should be processed before integration into the BIM environment. It was emphasized that the creation of a new subschema for surveying engineering in the domain layer of the IFC schema should be developed, which would in turn make engineering surveyors as part of the open BIM community. However, Clemen and Gründig (2006) did not provide an approach as to how 3D surveying data elements and indoor cadastral data could be meshed into and accommodated within the spatial data model of the IFC standard.

3.5.2 IFC for Property Valuation

The valuation of properties differs from one jurisdiction to another (Isikdag et al. 2014). 3D digital data environments, such as BIM, could potentially enhance valuation and taxation practices in the future; however, this improvement differs from one jurisdiction to another, since each jurisdiction has its own requirements for valuating properties. Property valuation is mainly predicated on physical world aspects (geometry and material of the building element) and legal/virtual aspects (RRR information and zoning of legal spaces). Therefore, Isikdag et al. (2014) investigated linking 3D cadastral (legal) models with 3D semantically enriched physical building models such as IFC, arguing that this could potentially improve property valuation practices. Current practices for determining property values and taxes were investigated in different countries including Turkey, the UK, the United States, the Netherlands, and Germany. They proposed that integrating 3D RRR information with 3D physical building models in either IFC or CityGML format could potentially provide significant benefits for the valuation and taxation of properties in these countries.

In further research by Isikdag et al. (2015), data requirements for property valuation were identified for each of the aforementioned countries. These data requirements were classified into two broad categories:

- Land lots: 2D spatial data required for valuation of land lots include 2D boundaries of land lots, 2D permit boundaries (such as building or farming), information for providing accessibility and utility services to a land lot, and vegetation information for protecting land lots (i.e. restrictions). Additionally, 3D spatial data requirements include the 3D land parcel with possible additional information about quality and usability, 3D volume space of permit boundaries, and 3D limits of trees, bushes, or fences in some jurisdictions.
- Buildings: 2D spatial data requirements for buildings include 2D floorplans (ground area and floor areas), orientation, and view of the building. In addition, 3D spatial data requirements include a detailed 3D indoor building model (LoD4+) for estimating value and cost, separate building parts, and a 3D model of the building and its neighborhood with realistic pictures such as *streetview* images.

One major finding reported by Isikdag et al. (2015) was that the enrichment of 3D digital BIM models with *streetview* images would provide more detailed information about the quality of the environment surrounding the building, which leads to improving the representation of the valuation information associated with apartment units of the building.

3.5.3 Cadastral Extension of UBM

The capability of the IFC standard and CityGML standards for implementing 3D cadastral systems was investigated by El-Mekawy and Östman (2012, 2015). They proposed that Unified Building Models (UBM) can be extended with four types of boundary surfaces which are required to represent aboveground and underground RRR spaces in the context of the Swedish jurisdiction. UBM is a reference data model for bi-directional mapping between IFC and CityGML standards, and it facilitates the exchange of 3D spatial information between these standards (El-Mekawy 2010). The proposed boundary surfaces in the cadastral extension of UBM are "Building Elements Surfaces," "Digging Surfaces," "Protecting Area Surfaces," and "Real Estate Boundary Surfaces" (El-Mekawy and Östman 2015). "Building Elements Surfaces" refer to the legal boundaries associated with building elements such as slabs and walls. "Digging Surfaces" are virtually defined legal boundaries for underground infrastructure, such as tunnels and underground parking areas. "Protecting Area Surfaces" are utilized to delineate the spatial extent of restrictions such as easements. "Real Estate Boundary Surfaces" are delineated based on traditional cadastral survey methods, in which border points and border lines constitute a cadastral object.

To showcase the feasibility of the proposed cadastral extension of UBM, it was implemented for a real case study, a hospital building, in an ArcGIS environment where various types of RRR spaces defined by the proposed boundary types were visualized (El-Mekawy and Östman 2012).

In a further study by El-Mekawy, Paasch and Paulsson (2014), the authors highlighted possible solutions, as well as associated challenges, for adopting BIM in urban land administration. They argued that BIM would potentially enhance the procedures associated with the formation, registration, and visualization of RRR information defined in the Swedish jurisdiction, particularly as the data environment of the BIM model does not currently include RRR information. Moreover, it was highlighted that with a Swedish property formation process, subdividing or consolidating land parcels or properties is mainly predicated on using digital building information supplied by construction companies in the form of BIM or CAD files. Nevertheless, there is no further use of this digital information after completion of the property formation process. To mitigate these challenges, the proposed cadastral extension of UBM was suggested as one possible solution to facilitate the interaction between BIM and 3D property formation. This indicates that the enrichment of UBM- or BIM-based data models with 3D RRR data elements could bring new opportunities to enable further collaboration between land administration and BIM domains.

Finally, El-Mekawy, Paasch, and Paulsson (2015) examined IFC, CityGML, and Land Administration Domain Model (LADM) standards in terms of modeling various types of rights, restrictions, and responsibilities (RRRs) associated with 3D properties in Sweden. The study mainly focused on presenting a theoretical framework to provide the basis for the interaction between 3D physical models, such as IFC and CityGML, and 3D legal models such as LADM. They found that both IFC and CityGML standards do not support the legal aspects of 3D properties such as boundaries and RRR information, whereas LADM provides a framework for only modeling legal spaces. Therefore, they found that the proposed extension to UBM would efficiently model legal aspects of 3D properties in comparison with those standards, particularly in complex structures. Another finding was that a full 3D cadastral system should logically incorporate both the legal and physical dimensions of 3D properties in urban built environments, which is still a remaining research question that needs to be addressed.

Although the proposed cadastral extension of UBM was a good contribution to using BIM models for urban land administration purposes, a few issues associated with this extension are:

- The spatial relationships between boundaries and legal spaces are not modeled in this extension. For instance, there is no support for how "Building Elements Surfaces" might reference walls or ceilings bounding a private apartment unit. This limitation applies to those jurisdictions (such as Victoria) that mainly rely on building elements to define legal boundaries.
- The extended cadastral UBM only represents how legal boundaries can be defined and does not model other main cadastral entities, which include information describing interest holders and legal documents, such as titles or deeds.
- The proposed UBM extension is very restricted and does not specify the detailed semantic information for 3D RRR spaces. For example, there is no information about liabilities and entitlements of private properties.
- Cadastral extension of UBM does not distinguish various legal arrangements from each other. For example, "UBMSpace" is the potential entity for modeling legal spaces and this entity does not provide more semantic information to distinguish a private property from common property space. Furthermore, UBM does not provide the capability to arrange different parts of a legal interest in a spatial structure. As an example, various communal spaces such as corridors, lobbies, and elevators typically constitute the common property legal interest of a multi-level building, but there is no mechanism in the cadastral extension of UBM to compose instances of the "UBMSpace" entity to form the whole spatial structure of common properties.

3.5.4 UrbanIT Project

The urbanIT integration framework was another relevant research project which focused on the development and implementation of an integrated urban information model by coupling indoor spatial information sourced from BIM with outdoor spatial information derived from geographic information systems (GIS) (Plume and Mitchell 2011). This framework would potentially provide the basis for effective management of spatial information, which would, in turn, enhance current processes for planning, designing, and operating urban built environments.

In particular, the core of the urbanIT project was a proposed extension to the IFC standard for managing cadastral data inside buildings, as well as in land parcels on the site of buildings (Barton et al. 2010). The proposed extension considered "IfcSpace" and "IfcZone" entities for modeling legal spaces were defined within multi-story buildings. In addition, the "IfcSite" entity was also considered for modeling legal boundaries of 2D land parcels on a construction site. To model cadastral attributes, the "IfcSite" entity was enriched with a new property set (*Pset_CadastreCommon*). This property set comprised four attributes, namely:

- *CadastreType* represents the legal interests associated with the land parcel such as road, lot, railway, water feature and so on.
- *CadastreSubType* defines the subtypes of a specific legal interest. For instance, lots can be specialized into strata and stratum lot, standard lot, and standard part lot.
- *LandValuation* provides the predicted value of assets, which could be obtained based on various valuation methods such as Fair Valuation and Market Valuation
- *Tenure* describes who owns or holds interests in the land.

Although this extension was a good trial for representing cadastral spaces within buildings, three main issues relating to the cadastral part of urbanIT were identified:

- It did not investigate how legal boundaries, particularly the range of cadastral boundaries defined inside high-rise and complex buildings, could be mapped onto their corresponding IFC entities.
- The concept of zones (IfcZone) was used for modeling common property areas, in which only cognitive volumetric spaces (IfcSpace) are used to model constituting parts of common property. However, in the notation section of subdivision plans, it is sometimes indicated that physical structures, such as walls and ceilings, pipes inside walls, and cables inside walls, are also deemed to be part of common properties. This implies that the "IfcZone" concept is unlikely to adequately manage spatial arrangements of common properties in some circumstances.
- Other important cadastral attributes, such as lot entitlements and liabilities, were not assigned as property sets for individually owned lots.

3.5.5 BIM-Based Registration of Stratified Properties in the Netherlands

BIM was used as a data source for 3D digital registration in the context of the Dutch jurisdiction. Stoter et al. (2017) described the procedures for registering stratified property rights in two case studies in 3D digital environments. The first case study of this project was based on a structural combination of the city hall and underground railway station in the city of Delft. The second case study was a complex development comprising a hotel, a residential building, and an underground car park in the northern area of Amsterdam. According to these case studies, a workflow was developed to register stratified properties using 3D PDF documents. This workflow includes two data source options for delineating the spatial extent of 3D legal spaces: sourcing 3D data from 3D surveying practices (such as laser scanning) or adopting architectural BIM models. The former approach is suitable for existing developments while the latter one can be used for new developments.

The second approach was only used for delineating and registering 3D legal spaces in the Dutch jurisdiction (Stoter et al. 2017). In this approach, the architectural BIM model of the development was transformed to 3D geometries defining the spatial extent of volumetric legal spaces. These legal spaces were defined by considering three factors: (1) the design model of the development; (2) the existing registered deed associated with the 2D cadastral map of the development; and (3) seeking input from relevant stakeholders. In the next step, some validation checks were conducted to make sure that each legal space was volumetrically closed, with no gaps or overlaps between any two legal spaces. In addition, the consistency of 3D legal spaces with their corresponding physical elements inside the design model was checked. After these checks, a 2D cadastral map within the vicinity of the development was integrated with the 3D digital model of stratified legal spaces. This 2D cadastral map provides contextual map information, such as a north-arrow, legend, and geospatial coordinates. In the final step, 3D PDF documents, which comprise both the 3D digital model and the 2D cadastral map, were created and registered as a legal document in the land registry organization. Additionally, the 3D digital data used for delineating the spatial extent of volumetric legal spaces is also maintained in the Dutch Kadaster for the purpose of future updates and modifications.

In both case studies, using the BIM-based approach for registering vertically stratified properties has brought significant benefits in comparison with the traditional 2D-based methods. One advantage was the provision of clearly represented multi-level property rights in the 3D digital environment of BIM. In addition, 3D legal spaces can be represented in one integrated data environment, rather than being projected into fragmented 2D floorplans and cross-section diagrams, which consequently reduces the mental exercise required for understanding the spatial extent of these legal spaces. Another benefit was the cost effectiveness of registering complex ownership situations in 3D digital environments. One reason is that the difficult task of delineating complex ownership boundaries using 2D-based descriptions can be eliminated. Moreover, there is no need to conduct surveying for ground parcels within complex developments.

Stoter et al. (2017) also identified that changes are required in current legislation and procedures in order to realize 3D digital registration systems. Existing legal frameworks in the cadastral and land acts should be modified. It would take many years to amend legislation in support of full 3D digital cadastral registration. Therefore, 3D digital representation of legal spaces can come into effect if juridical and organizational issues are addressed adequately.

3.5.6 Connecting BIM and Legal Data Models

Connecting legal data models with physical data models is a new area of research and development in urban land administration. A wide range of approaches is proposed for this integration. Some solutions suggest the integration can be done only for visualization purposes, while others propose that specific relationships between legal spaces and physical components should be defined when it is required to do so. The latter approach includes various combinations of major legal data models, which are LADM, ePlan, and LandInfra, and major physical data models including IFC, CityGML, and IndoorGML. A comprehensive assessment of these combinations for managing legal interests and boundaries defined within multi-owned developments in Victoria, Australia has been presented in Rajabifard, Atazadeh, and Kalantari 2018. As this book focuses on the BIM environment or IFC standard, various approaches adopted for connecting BIM and legal data models will be presented in the following subsections.

3.5.6.1 IFC and LADM

Two pathways have been suggested for connecting IFC and LADM standards (Atazadeh, Rajabifard, and Kalantari 2018). These pathways are shown diagrammatically in Figure 3.8. The first pathway is to extend LADM with physical objects adopted from the IFC standard. In this approach, it is suggested that physical objects should be defined explicitly inside the future version of the LADM standard. In addition, this new potential version of the LADM standard should provide the ability to define optional semantic relationships between physical elements and legal entities. This includes defining relationships between legal boundary surfaces (defined by LA_BoundaryFace in LADM) and physical elements (defined by IfcElement in IFC), as well as defining relationships which provide the possibility of assigning a physical element as a constituent part of a basic administrative unit (defined by LA_BAUnit in LADM).

The main benefit of extending LADM with physical objects is to encourage those jurisdictions, such as Australian states, which rely on physical objects to

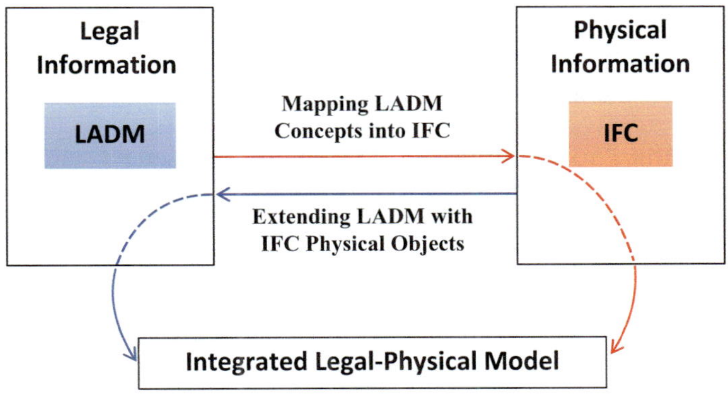

FIGURE 3.8
Proposed pathways for developing an integrated model based on IFC and LADM.

define the legal extent of property rights, to adopt an LADM-based approach for 3D digital management of spatially complex legal spaces. However, this pathway seems very unlikely and less feasible, since LADM is a conceptual model and the proposed integrated legal–physical model would not be able to respond to specifics of legal and physical relationships defined in the context of all jurisdictions around the world.

The second pathway is to perform a complete mapping of LADM concepts to the implementation schema of the IFC standard. In light of this pathway, Atazadeh, Rajabifard and Kalantari (2018) and Oldfield et al. (2017) proposed suitable mechanisms for encoding LADM-based legal concepts into the current data structure of the IFC standard.

Oldfield et al. (2017) suggested that space objects (IfcSpace) and grouping these spaces as legal zones (IfcZone) would underpin the basis for utilizing BIM models in the 3D cadaster. It was also proposed that boundaries of legal spaces could be modeled by "IfcRelSpaceBoundary." Using a lakeside restaurant as a case study, two main challenges associated with adopting the IFC standard for 3D cadastral purposes were identified, namely the manual extraction of the spatial extent of 3D legal interests from BIM models, and modeling ambulatory boundaries defined by changes in the water levels of the lake located around the restaurant. For modeling the ambulatory boundary, the study proposed to enrich "IfcRelspaceBoundary" with a semantic value for ambulatory boundaries. However, this approach may not be a complete method for modeling ambulatory boundaries. Since these boundaries are defined by referencing rivers or lakes, which are currently modeled by the "IfcGeogrpahiclElement" entity, it could be better to point out that "IfcRelSpaceBoundary" should refer to "IfcGeographicElement" via its *RelatedBuildingElement* attribute (see Chapter 4).

Atazadeh, Rajabifard and Kalantari (2018) proposed a more comprehensive approach for connecting LADM and IFC standards. Their suggested method comprises two major steps: identifying suitable IFC entities for mapping an

LADM concept itself, and modeling attributes of each LADM concept using property sets. How each concept can be mapped into the IFC standard will be explained in the following:

- Spatial units: Spatial units can be visually represented using a wide range of geometries in different dimensions. In the context of 3D digital urban land administration, major forms of spatial units are land parcels and indoor and outdoor legal spaces. Table 3.1 provides those IFC entities which are suitable for representing each type of spatial unit.
- Boundary concepts: There are two boundary concepts in LADM, namely, LA_BoundaryFaceString for string boundaries and LA_BoundaryFace for face boundaries. These concepts refer to both the topology and geometry of string and face boundaries. Therefore, the LA_BoundaryFaceString concept should be implemented by the IfcEdeCurve entity which supports both geometry and topology of boundary lines. For LA_BoundaryFace, it is suggested that IfcFaceSurface is adopted for implementing both the geometry and topology of boundary faces.
- Basic administrative units and RRR concepts: There are no suitable IFC entities for mapping basic administrative units (LA_BAUnit) and RRR concepts (LA_RRR and its subclasses LA_Right, LA_Restriction, and LA_Responsibility). Using property sets was considered as a possible approach to define attributes of these concepts inside the IFC standard.
- Party concepts: In order to map party concepts (LA_Party, LA_GroupParty, and LA_PartyMember) into the IFC standard, the IfcActor entity can be considered as a suitable option. This entity refers to other IFC entities defined in the IfcActorResource subschema for specifying party roles (via IfcActorRole entity) and determining party types, including person (IfcPerson), organization (IfcOrganization), and person on behalf of an organization (IfcPersonAndOrganization).

TABLE 3.1

Modeling Spatial Units Within the IFC Standard

Type of Spatial Unit		Counterpart Entities in IFC
Land parcel	Single	IfcSite
	Multiple	IfcSpatialZone
Indoor legal space	Single	IfcSpace
	Multiple	IfcZone, IfcSpatialZone
Outdoor legal space	Single	IfcExternalSpatialElement
	Multiple	IfcSpatialZone

- Administrative sources: In LADM, LA_AdministrativeSource is the concept adopted for managing legal documents. To map this concept into the IFC standard, there are two suitable entities, namely IfcDocumentReference and IfcDocumentInformation. IfcDocumentReference provides the location of a document via its *Location* attribute. IfcDocumentInformation entity also provides additional attributes to provide the metadata about a document. This metadata includes the description of content, purpose, intended use, revision information, editor, creation time, electronic format, confidentiality and so on.

Mapping LADM concepts into the IFC standard seems a more promising solution for the integrated physical–legal model compared to the first pathway. This approach can bring significant advantages, such as linking LADM data with other lifecycle data about buildings. Land administration actors, such as surveying professionals, can better communicate with other land development stakeholders. However, there are a number of existing challenges to realize the connection between LADM and IFC. These challenges include developing cooperative interactions between standardization specialists involved in these standards, and providing support for testing and implementing the model in software applications.

3.5.6.2 BIM and the ePlan Model

The interaction between BIM and the ePlan model was investigated by a 3D digital cadaster research team in the land registry organization of Victoria (Shojaei et al. 2016). A 3D ePlan prototype system was designed and implemented for a complex multi-story development. In this prototype system, 3D digital was sourced from the BIM environment. More specifically, the 3D digital BIM model comprising both physical elements and legal spaces was constructed inside the Autodesk Revit which is a well-known BIM authoring tool. In the next step, a series of data conversions were performed to transform the 3D digital data from the BIM environment to the valid LandXML format which is compliant with the ePlan protocol.

The results of this project indicated that the full representation of the spatial extent of complex ownership structures, such as common properties, can be challenging for the ePlan protocol. Another challenge was associated with the smooth representation of the geometry of curved boundaries which are defined occasionally in building subdivision plans. The conversion of data from IFC into the LandXML format resulted in flat faces when approximating the curved surfaces created in the BIM environment. Although this approximation would be adequate for registering curved-shape legal spaces, it may affect the computations and analysis associated with the area and volume quantities of these legal spaces.

3.5.6.3 BIM and LandInfra

The LandInfra standard was developed in line with the IFC alignment project. In this project, the IFC standard has been extended for modeling the lifecycle information for infrastructure facilities such as roads, bridges, and tunnels. LandInfra and IFC alignment projects were joint efforts between OGC and BuildingSMART organizations to facilitate data interoperability between BIM and GIS environments (Scarponcini et al. 2016). Relevant to the purpose of urban land administration, the connection can mainly be observed in the equivalence between the "CondominiumScheme" class in LandInfra and the "IfcSpatialZone" entity in IFC. A condominium ownership is structured based on a collection of building parts. The concept of building parts (BuildingPart) in LandInfra is similar to the concept of spatial elements (IfcSpatialElement) as defined in the IFC standard.

3.5.7 IDM-Based Workflow for Cadastral Registration

Oldfield et al. (2016) developed a collaborative IDM-based workflow for cadastral registration. The workflow described how cadastral data requirements could be efficiently communicated between project initiators and authorities, which would, in turn, facilitate procedures for obtaining legal spaces from BIM models. One issue with the proposed workflow is that it did not distinguish specific acting roles performed by authorities. For example, responsible authorities (or city councils) were not differentiated from referral authorities, such as utility companies.

In a further study, Oldfield et al. (2018) looked at coordinating the process for obtaining building permits using IDM-based workflows. They mapped the transactions between the project initiator and municipality in two phases: pre-construction and post-construction. In these transactions, it is assumed that IFC files comprising spatial units and RRR information along with survey measurements are sent for building permit approval. In the pre-construction phase, as-designed BIM models are used for granting or declining the building permit. In the post-construction phase, as-built BIM models are examined to finalize a building permit application. Using XML-based tools was also suggested for automating multi-actor collaboration and communication during the process of a building permit application.

3.5.8 As-Built BIM Models for Underground Property Management

An implementation framework encompassing indoor mapping and as-built BIM models was proposed to manage RRR information associated with 3D properties located underground in the context of the Korean jurisdiction (Kim et al. 2015b). The proposed implementation includes four main steps. First, geometric information relating to the infrastructure facilities was modeled using data sourced from terrestrial laser scanners. Subsequently, an as-built BIM model according to the results of the geometric modeling was constructed. This was followed by defining reference points in order to verify the accuracy of the as-built BIM model. Finally, three approaches were suggested as possible solutions for representing and visualizing 3D underground properties. These approaches include: using isometric views to represent 3D properties, using a 2D land parcel represented with 2D footprints of 3D underground properties, or using 3D (or 2.5D) surface and 3D underground cadastral maps. It is claimed by Kim et al. (2015b) that their framework can register RRR information associated with underground legal interests; however, this framework only considers the purely geometric aspects of underground cadastral objects and does not provide semantic mechanisms to assign RRR information for each legal interest. In addition, there is no explanation for how the geometry of physical objects obtained from indoor mapping can be used to delineate structural boundaries.

3.6 Concluding Remarks

The beginning of this chapter introduced the fundamental concepts of the BIM environment. These concepts include process and product perspectives on BIM, maturity levels of BIM in the AEC industry, and multi-dimensional BIM models. A discussion on the significant role of open BIM standards in mitigating challenges associated with interoperability and exchange of BIM data in the AEC industry has been highlighted.

In open BIM environments, the IFC standard is a comprehensive open data model used for managing the entire lifecycle information for buildings. This standard has four main layers, namely the resource layer, the core layer, the interoperability layer, and the domain layer. The IFC standard provides the data schema for 3D visualization, structural analysis, energy analysis, solid models, physical and spatial connectivity relationships, scheduling-related information, cost estimation, various types of spatial reference systems, transformation between these systems, controlling contracts, monitoring changes, and much more. However, efficient implementation and use of the IFC standard in the AEC projects is heavily dependent on other open BIM standards, namely IDM, MVD, IFD, and BCF.

Those IFC entities that are important for urban land administration were also highlighted in this chapter. Moreover, the IFC standard and CityGML standard were also compared to highlight their differences and to discuss why IFC could be a better choice for urban land administration. Finally, relevant preliminary projects using BIM for land administration purposes were reviewed. Although these reviewed projects show different degrees of leveraging BIM for supporting land administration functions, none of them have comprehensively explored how the IFC standard can be extended to accommodate a wide range of legal data elements, which is critical for the adoption of BIM in the context of urban land administration. Thus, in the next chapter, the fundamental approach for meshing legal information together with an open BIM environment will be presented.

References

Agoston, Max K. 2005. *Computer Graphics and Geometric Modeling*. Vol. 1. London, UK: Springer.

Allen, George. 1984. "An Introduction to Solid Modelling." *Computers and Graphics* 8 (4):439–447.

Atazadeh, Behnam, Mohsen Kalantari, and Abbas Rajabifard. 2016. "Comparing Three Types of BIM-Based Models for Managing 3D Ownership Interests in Multi-Level Buildings." In: *5th International FIG 3D Cadastre Workshop*, 183–198, 18–20 October 2016, Athens, Greece. International Federation of Surveyors (FIG). Available at: www.gdmc.nl/3DCadastres/workshop2016/programme/Workshop2016_10.pdf.

Atazadeh, Behnam, Mohsen Kalantari, Abbas Rajabifard, Serene Ho, and Tom Champion. 2017. "Extending a BIM-Based Data Model to Support 3D Digital Management of Complex Ownership Spaces." *International Journal of Geographical Information Science* 31 (3):499–522.

Atazadeh, Behnam, Abbas Rajabifard, and Mohsen Kalantari. 2018. "Connecting LADM and IFC Standards – Pathways towards an Integrated Legal-Physical Model." In: Christiaan Lemmen, and Peter van Oosterom (eds.), *7th International FIG Workshop on the Land Administration Domain Model*, 89–102, 12–13 April 2018, Zagreb, Croatia. Available at: https://wiki.tudelft.nl/pub/Research/ISO19152/LADM2018Workshop/06-07_LADM_2018.pdf.

Barton, Jack, David Marchant, John Mitchell, Jim Plume, and Peter Rickwood. 2010. *A Note on Cadastre: UrbanIT Research Project*. City Futures Research Centre.

Bazjanac, V. 2004. "Virtual Building Environments (VBE)–Applying Information Modeling to Buildings." In: *Proceedings of the European Conference on Product and Process Modelling in the Building and the Construction Industry (ECPPM)*, 41–48, 8–10 September 2004, Istanbul, Turkey.

Becker, Thomas, Claus Nagel, and Thomas H. Kolbe. 2013. "Semantic 3D Modeling of Multi-Utility Networks in Cities for Analysis and 3D Visualization." In: Jacynthe Pouliot, Sylvie Daniel, Frédéric Hubert, and Alborz Zamyadi (eds.), *Progress and New Trends in 3D Geoinformation Sciences SE-3*, 41–62. Lecture Notes in Geoinformation and Cartography. Berlin/Heidelberg: Springer. https://doi.org/10.1007/978-3-642-29793-9_3.

Bell, Håvard, and Lars Bjørkhaug. 2006. "A BuildingSMART Ontology." In: *Proceedings of the 2006 ECPPM Conference*, 185–90, 13–15 September 2005, Valencia, Spain.

British Standards Institution. 2013. *PAS 1192–2: 2013: Specification for Information Management for the Capital/Delivery Phase of Construction Projects Using Building Information Modelling*. London, UK: BSI.

BuildingSMART. 2013a. *IFC Control Extension*. Available at: http://www.buildingsmart-tech.org/ifc/IFC4/final/html/schema/ifccontrolextension/content.htm.

BuildingSMART. 2013b. *IFC Process Extension*. Available at: http://www.buildingsmart-tech.org/ifc/IFC4/final/html/schema/ifcprocessextension/content.htm.

BuildingSMART. 2013c. *IFC Product Extension*. Available at: http://www.buildingsmart-tech.org/ifc/IFC4/final/html/schema/ifcproductextension/content.htm.

BuildingSMART. 2013d. *Local Placements in IFC*. Available at: http://www.buildingsmart-tech.org/ifc/IFC4/final/html/schema/ifcgeometricconstraintresource/lexical/ifclocalplacement.htm.

BuildingSMART. 2013e. *Object Placements in IFC*. Available at: http://www.buildingsmart-tech.org/ifc/IFC4/final/html/schema/ifcgeometricconstraintresource/lexical/ifcobjectplacement.htm.

BuildingSMART. 2013f. *Resource Definition Data Schemas*. Available at: http://www.buildingsmart-tech.org/ifc/IFC2x4/rc2/html/schema/chapter-8.htm.

BuildingSMART. 2013g. *Root Class of IFC Standard*. Available at: http://www.buildingsmart-tech.org/ifc/IFC4/final/html/schema/ifckernel/lexical/ifcroot.htm.

BuildingSMART. 2013h. *Shape Representations in IFC Standard*. Available at:http://www.buildingsmart-tech.org/ifc/IFC4/final/html/schema/ifcrepresentationresource/lexical/ifcshaperepresentation.htm.

BuildingSMART. 2013i. *Shared Element Data Schemas*. Available at: http://www.buildingsmart-tech.org/ifc/IFC4/final/html/schema/chapter-6.htm.

BuildingSMART. 2013j. *Site Elements in IFC*. Available at: http://www.buildingsmart-tech.org/ifc/IFC4/final/html/schema/ifcproductextension/lexical/ifcsite.htm.

BuildingSMART. 2017. *Open BIM Standards*. Available at: https://www.buildingsmart.org/about/what-is-openbim/ifc-introduction/.

Centre for Digital Built Britain. 2011. "A Report for the Government Construction Client Group Building Information Modelling (BIM) Working Party Strategy Paper." London, UK.

Centre for Digital Built Britain. 2015. *Digital Built Britain Level 3 Building Information Modelling Strategic Plan*. London, UK.

Clemen, Christian and L. Gründig. 2006. "The Industry Foundation Classes (IFC)–Ready for Indoor Cadastre?" In: *Proceedings of XXIII International FIG Congress*. Vol. 18, 8–13 October 2006, Munich, Germany

Eastman, Charles M. 1999. *Building Product Models: Computer Environments, Supporting Design and Construction*. CRC Press.

Eastman, Charles M., Paul Teicholz, Rafael Sacks, and Kathleen Liston. 2011. *BIM Handbook: A Guide to Building Information Modeling for Owners, Managers, Designers, Engineers and Contractors*. John Wiley & Sons.

Ekberg, Fredrik. 2007. *An Approach for Representing Complex 3D Objects in GIS Applied to 3D Properties*. Master's thesis, University of Gävle, Sweden.

El-Mekawy, Mohamed. 2010. "Integrating BIM and GIS for 3D City Modelling: The Case of IFC and CityGML." *Trita-SOM. KTH, Geoinformatics: KTH*. Available at: http://urn.kb.se/resolve?urn=urn:nbn:se:kth:diva-28899.

El-Mekawy, Mohamed and Anders Östman. 2015. "A Unified Building Model for a Real 3D Cadastral System." In: Carlos Nunes Silva (ed.), *Emerging Issues, Challenges, and Opportunities in Urban E-Planning*, 252. IGI Global.

El-Mekawy, Mohamed and Anders Östman. 2012. "Feasibility of Building Information Models for 3D Cadastre in Unified City Models." *International Journal of E-Planning Research* 1 (4):35–58.

El-Mekawy, Mohamed, Anders Östman, and Ihab Hijazi. 2012. "A Unified Building Model for 3D Urban GIS." *ISPRS International Journal of Geo-Information* 1 (2):120–145.

El-Mekawy, Mohamed, Jesper Paasch, and Jenny Paulsson. 2014. "Integration of 3D Cadastre, 3D Property Formation and BIM in Sweden." In: *4th International Workshop on 3D Cadastres*, 17–34, 9–11 November 2014, Dubai, United Arab Emirates.

El-Mekawy, Mohamed, Jesper M. Paasch, and Jenny Paulsson. 2015. "Integration of Legal Aspects in 3D Cadastral Systems." *International Journal of E-Planning Research* 4 (3):47–71.

Frédéricque, Benoit, Keith Raymond, and Kees Van Prooijen. 2011. "3D GIS as Applied to Cadastre–A Benchmark of Today's Capabilities." In: *FIG Working Week*, 18–22 May 2011, Marrakech, Morocco.

Giel, Brittany Kathleen. 2009. "Return on Investment Analysis of Building Information Modeling in Construction." Master's thesis, University of Florida.

Goedert, James D. and Pavan Meadati. 2008. "Integrating Construction Process Documentation into Building Information Modeling." *Journal of Construction Engineering and Management* 134 (7):509–516. American Society of Civil Engineers.

Gröger, Gerhard and Lutz Plümer. 2012. "CityGML – Interoperable Semantic 3D City Models." *ISPRS Journal of Photogrammetry and Remote Sensing* 71:12–33.

Hammoudi, Karim. 2011. "Contributions to the 3D City Modeling: 3D Polyhedral Building Model Reconstruction from Aerial Images and 3D Facade Modeling from Terrestrial 3D Point Cloud and Images." Ph.D. Thesis, Université Paris-Est, France.

Ho, Serene, Abbas Rajabifard, and Mohsen Kalantari. 2015. "'Invisible'Constraints on 3D Innovation in Land Administration: A Case Study on the City of Melbourne." *Land Use Policy* 42:412–425.

Isikdag, Umit. 2015. "Building Information Models: An Introduction." In: *Enhanced Building Information Models – Using IoT Services and Integration Patterns*, 1–12. SpringerBriefs in Computer Science. Springer International Publishing. doi:10.1007/978-3-319-21825-0_1.

Isikdag, Umit, Ghassan Aouad, Jason Underwood, and Song Wu. 2007. "Building Information Models: A Review on Storage and Exchange Mechanisms." In: *24th CIB W78 Conference*, 26–29 June 2007, Maribor, Slovenia.

Isikdag, Umit, Mike Horhammer, Sisi Zlatanova, R. Kathmann, and P. J. M. Van Oosterom. 2014. "Semantically Rich 3D Building and Cadastral Models for Valuation." In: *Proceedings 4th International Workshop on 3D Cadastres*, 9–11 November 2014, Dubai, United Arab Emirates. International Federation of Surveyors (FIG).

Isikdag, Umit, Mike Horhammer, Sisi Zlatanova, R. Kathmann, and Peter van Oosterom. 2015. "Utilizing 3D Building and 3D Cadastre Geometries for Better Valuation of Existing Real Estate." In: *FIG Working Week 2015: From the Wisdom of the Ages to the Challenges of the Modern World*, 1–18, 17–21 May 2015, Sofia, Bulgaria.

Isikdag, Umit, Jason Underwood, and Ghassan Aouad. 2008. "An Investigation into the Applicability of Building Information Models in Geospatial Environment in Support of Site Selection and Fire Response Management Processes." *Advanced Engineering Informatics* 22 (4):504–519.

Isikdag, Umit, S. Zlatanova, and J. Underwood. 2011. "An Opportunity Analysis on the Future Role of BIMs in Urban Data Management." In: *Urban and Regional Data Management–UDMS Annual*, 25–36.

ISO. 2004. "Virtual Reality Modeling Language (VRML) ISO/IEC 14772-1:1997 and ISO/IEC 14772-2:2004." International Standard ISO/IEC. International Organization for Standardization.

ISO. 2005. ISO/IEC FCD 19775 –1r1:200x Information Technology—Computer Graphics, Image Processing and Environmental Representation—Extensible 3D (X3D)—Part 1: Architecture and Base Components. International Organization for Standardization.

ISO/PAS17506. 2012. "Industrial Automation Systems and Integration—COLLADA Digital Asset Schema Specification for 3D Visualization of Industrial Data." International Organization for Standardization.

ISO10303-41. 2005. "Product Data Representation and Exchange." International Organization for Standardization, Subcommittee.

ISO16739. 2013. "Industry Foundation Classes (IFC) for Data Sharing in the Construction and Facility Management Industries." International Organization for Standardization.

Jarroush, Jad and Gilad Even-Tzur. 2004. "Constructive Solid Geometry as the Basis of 3D Future Cadastre." In: *FIG Working Week*, Vol. 2004, 22–27 May 2004, Athens, Greece.

Kim, Jung, Jonghoon Kim, Martin Fischer, and Ryan Orr. 2015a. "BIM-Based Decision-Support Method for Master Planning of Sustainable Large-Scale Developments." *Automation in Construction* 58:95–108.

Kim, Sangmin, Jeonghyun Kim, Jaehoon Jung, and Joon Heo. 2015b. "Development of a 3D Underground Cadastral System with Indoor Mapping for As-Built BIM: The Case Study of Gangnam Subway Station in Korea." *Sensors* 15 (12):30870–30893.

Kolbe, Thomas. 2012. "BIM, CityGML, and Related Standardization." In: *Digital Landscape Architecture Conference 2012*, 31 May–2 June 2012, Bernburg/Dessau, Germany.

Kolbe, Thomas, Gerhard König, and Claus Nagel. 2011. *Advances in 3D Geo-Information Sciences*. Springer Science & Business Media.

Kolbe, Thomas, Claus Nagel, and Alexandra Stadler. 2009. "CityGML–OGC Standard for Photogrammetry." *Photogrammetric Week* 9:265–277.

Lee, Ghang, Harrison Kwangho Park, and Jongsung Won. 2012. "D3 City Project—Economic Impact of BIM-Assisted Design Validation." *Automation in Construction* 22:577–586.

Liebich, T. 2009. *IFC 2x Edition 3. Model Implementation Guide. Version 2.0*. AEC3 Ltd.

Mantyla, Martti and Reijo Sulonen. 1982. "GWB: A Solid Modeler with Euler Operators." *IEEE Computer Graphics and Applications* 2 (7):17–31.

Nagel, Claus. 2014. "Spatio-Semantic Modelling of Indoor Environments for Indoor Navigation." Available at: https://opus4.kobv.de/opus4-tuberlin/frontdoor/index/index/docId/5537.

Nagel, Claus, Alexandra Stadler, and Thomas Kolbe. 2009. "Conceptual Requirements for the Automatic Reconstruction of Building Information Models from Uninterpreted 3D Models." In: *Proceedings of the International Archives of Photogrammetry, Remote Sensing and Spatial Information Sciences*, 34 (XXX):46–53.

NBIMS. 2012. "National BIM Standard - United States™ Version 2." http://www.nationalbimstandard.org/nbims-us-v2/pdf/NBIMS-US2_aB.pdf.

Oldfield, Jennifer, Ronald Bergs, Peter van Oosterom, Thomas Krijnen, and Miguel Galano. 2018. "3D Cadastral Lifecycle: An Information Delivery Manual ISO 29481 for 3D Data Extraction from the Building Permit Application Process." In: Christiaan Lemmen, and Peter van Oosterom (eds.), *7th International FIG Workshop on the Land Administration Domain Model*, 12–13 April 2018, Zagreb, Croatia. Available at: https://wiki.tudelft.nl/pub/Research/ISO19152/LADM2018Workshop/11-21_LADM_2018.pdf.

Oldfield, Jennifer, Peter van Oosterom, Jakob Beetz, and F. Thomas Krijnen. 2017. "Working with Open BIM Standards to Source Legal Spaces for a 3D Cadastre." *ISPRS International Journal of Geo-Information* 6 (11): 351.

Oldfield, Jennifer, Peter van Oosterom, Wilko Quak, Jeroen Van Der Veen, and Jakob Beetz. 2016. "Can Data from BIMs Be Used as Input for a 3D Cadastre?" In: Efi Dimopoulou and Peter van Oosterom (Eds.), *5th International FIG 3D Cadastre Workshop*, 199–214, 18–20 October 2016, Athens, Greece, International Federation of Surveyors (FIG). Available at: www.gdmc.nl/3DCadastres/workshop2016/programme/.

Pärn, E. A., D. J. Edwards, and M. C. P. Sing. 2017. "The Building Information Modelling Trajectory in Facilities Management: A Review." *Automation in Construction* 75:45–55. Elsevier.

Plume, Jim and John Mitchell. 2011. "An Urban Information Framework to Support Planning, Decision-Making & Urban Design." In: Pierre Leclercq, Ann Heylighen, and Geneviève Martin (Eds.), *14th International Conference on Computer Aided Architectural Design – Designing Together*, 653–66. Liège.

Popov, Vladimir, Saulius Mikalauskas, Darius Migilinskas, and Povilas Vainiūnas. 2006. "Complex Usage of 4D Information Modelling Concept for Building Design, Estimation, Scheduling and Determination of Effective Variant." *Ukio Technologinis Ir Ekonominis Vystymas* 12 (2):91–98. Taylor & Francis.

Pratt, Michael J. 2001. "Introduction to ISO 10303—The STEP Standard for Product Data Exchange." *Journal of Computing and Information Science in Engineering* 1 (1):102–103.

Rajabifard, Abbas, Behnam Atazadeh, and Mohsen Kalantari. 2018. "A Critical Evaluation of 3D Spatial Information Models for Managing Legal Arrangements of Multi Owned Developments in Victoria, Australia." *International Journal of Geographical Information Science* 32 (10):2098–2122.

Rossignac, Jaroslaw R. and Aristides A. G. Requicha. 1999. *Solid Modeling*. Georgia Institute of Technology. Available at: https://smartech.gatech.edu/handle/1853/3375.

Scarponcini, Paul, Hans-Christoph Gruler, Erik Stubkjær, Peter Axelsson, and Lars Wikstrom. 2016. "OGC® Land and Infrastructure Conceptual Model Standard (LandInfra)."

See, Richard and Jan Karlshoej. 2012. "An Integrated Process for Delivering IFC Based Data Exchange." Available at: www.standard.no/Global/PDF/ISO-TC59-SC13/N_287_Integrated_IDM-MVD_Process_for_IFC-formats.pdf.

Shojaei, Davood. 2014. "3D Cadastral Visualisation: Understanding Users' Requirements." Ph.D. thesis, University of Melbourne, Australia. Available at: http://hdl.handle.net/11343/45129.

Shojaei, Davood, Hamed Olfat, Abbas Rajabifard, Ashley Darvill, and Mark Briffa. 2016. "Assessment of the Australian Digital Cadastre Protocol (EPlan) in Terms of Supporting 3D Building Subdivisions." *Land Use Policy* 56:112–124.

Singh, Vishal, Ning Gu, and Xiangyu Wang. 2011. "A Theoretical Framework of a BIM-Based Multi-Disciplinary Collaboration Platform." *Automation in Construction* 20 (2):134–144.

Smith, Dana K. and Michael Tardif. 2009. *Building Information Modeling: A Strategic Implementation Guide for Architects, Engineers, Constructors, and Real Estate Asset Managers*. John Wiley & Sons.

Stadler, Alexandra and Thomas Kolbe. 2007. "Spatio-Semantic Coherence in the Integration of 3D City Models." In: *Proceedings of the 5th International Symposium on Spatial Data Quality*, 13–15 June 2007, Enschede, the Netherlands.

Stangeland, B. 2013. "BIM Collaboration Format (BCF)." *BuildingSMART International User Group*, Bakkmoen, Norway.

Stoter, Jantien, Hendrik Ploeger, Ruben Roes, Els van der Riet, Filip Biljecki, Hugo Ledoux, Dirco Kok, and Sangmin Kim. 2017. "Registration of Multi-Level Property Rights in 3D in the Netherlands: Two Cases and Next Steps in Further Implementation." *ISPRS International Journal of Geo-Information* 6 (6).

Succar, Bilal. 2009. "Building Information Modelling Framework: A Research and Delivery Foundation for Industry Stakeholders." *Automation in Construction* 18 (3):357–375.

Wikipedia. 2014. "Constructive Solid Geometry." Available at: https://en.wikipedia.org/wiki/Constructive_solid_geometry.

Wilson, Tim. 2008. "OGC Implementation Specification 07–147r2: OGC KML." In: *Open Geospatial Consortium Technical Report*.

Wilson, Tim, D. Burggraf, R. Lake, S. Patch, B. McClendon, M. Jones, M. Ashbridge, B. Hagemark, J. Wernecke, and C. Reed. 2007. "KML 2.2—An OGC Best Practice." In: *Open Geospatial Consortium, Document OGC07-113r1*.

Wix, Jeffrey and J. Karlshoej. 2010. "Information Delivery Manual: Guide to Components and Development Methods." *BuildingSMART International*. Available at: http://iug.buildingsmart.org/idms/development/IDMC_004_1_2.pdf.

4

BIM-Enabled 3D Digital Urban Land Administration—New Era in Managing Cities

4.1 Introduction

Digital urban land administration was described in Chapter 2 as a new paradigm for managing the legal and physical complexities of cities. Following on from this, Chapter 3 introduced BIM as a 3D digital environment providing a window on opportunities for the land administration domain. This chapter presents the new approach for colliding loosely coupled domains of BIM and urban land administration, which would lay the foundation for transforming current 2D-based land and property development practices into collaborative 3D digital environments.

BIM-enabled 3D digital urban land administration is underpinned by mapping the legal data elements onto the open IFC data model. One reason for choosing IFC in this book is that its data model provides a universal and open language for exchanging and sharing data in the BIM domain. Another reason is that IFC provides the most comprehensive data elements required for managing the lifecycle of building developments. This indicates that an extended version of the IFC standard can provide a reliable basis for developing and implementing a BIM-driven digital urban land administration. Therefore, the focus of this chapter is to describe the suitable mechanisms for using and extending the IFC standard with the legal data elements required for urban land administration. In Chapter 2, these elements were identified based on the data requirements for urban land administration. To recap from Section 2.4, these data elements include legal interests, legal boundaries, legal documents, land administration actors, and administrative data.

This chapter begins by first describing different alternatives for extending the IFC standard and the reasons for choosing the most appropriate extension method. Next, the major parts of the adopted extension method are described in detail. This is followed by an explanation of the details for modeling each legal data element within the extended version of the IFC standard. It should be noted that since legal data elements are highly dependent on the jurisdictional settings, the extension of the IFC standard presented in this chapter is based on those elements identified in the context of the Victorian jurisdiction. However, the chapter concludes by providing general and jurisdiction independent guidelines for realizing a BIM-based approach for urban land administration.

4.2 Alternatives for Extending IFC for Urban Land Administration

There are various methods of incorporating new data elements into the IFC standard. Although the IFC standard is primarily developed in the EXPRESS language, there is also an XML schema of this standard, which subsequently widens the options for enriching IFC with legal data elements. There are three major approaches for extension:

- Defining new subtypes in either the EXPRESS or XML schema: In this method, legal data elements of urban land administration are defined as subtypes for the appropriate IFC entities. For example, a legal space will be created by defining a new subtype, say "IfcLegalSpace," of the "IfcSpace" entity.
- Using available extension mechanisms in the EXPRESS schema: This approach uses extension mechanisms inside the IFC standard to model new concepts without the requirement to define new subtypes. For an example, the property sets concept can be used to attach the legal attributes to a space described by "IfcSpace," which subsequently makes it a legal space.
- Establishing an external linkage between the IFC standard and existing legal data models: This approach does not internally extend the IFC standard with legal data elements, but attempts to link open BIM datasets to the legal information maintained in another data repository.

Among the above-mentioned methods, the approach "using available extension mechanisms in the EXPRESS schema" is considered the appropriate mechanism for extending IFC in urban land administration. The reason for choosing this approach was that the BuildingSMART organization recommends it for incorporating domain-specific data elements into the IFC standard. In other words, it is in line with strategies that the BuildingSMART organization adopts for further developments of the IFC standard. In addition, it is a more optimal approach than defining new subclasses for the IFC standard, because new attributes and object types can be defined with minimal change in the current data structures of the IFC standard. This approach does not complicate the current IFC standard by adding new data elements. The main reason for not adopting options using the XML schema is that IFC was first developed based on the EXPRESS schema, while the XML schema is merely a replica. Moreover, the XML schema does not include all IFC entities defined in the EXPRESS schema, with some entities missing from the schema.

4.3 Adopted Extension Approach for Enabling IFC-Based Urban Land Administration

To build an extension of the IFC standard for urban land administration, two steps should be considered. First, suitable and relevant IFC entities to be extended should be identified. The IFC entities pertinent to urban land administration have been described in Section 3.4.8. In the second step, the selected extension approach should be applied to the relevant IFC entities. The adopted approach for extending IFC for urban land administration mainly relies on the concept of property set definitions. In addition, user-defined values inside the IFC standard are also adopted in this extension.

Keeping in line with the logic presented in the IFC standard, the relevant IFC entities will be further extended to model legal information with as minimal change as possible to the current data structure of IFC. Various EXPRESS-G diagrams will be used to facilitate a better understanding of the proposed entities and their corresponding relationships. Within these diagrams, the existing entities will be represented according to the coloring scheme used in the IFC standard. To understand these diagrams and their coloring scheme, readers are referred to Appendix A. To distinguish the proposed entities from the existing ones, they will be represented in red in all diagrams.

The following subsections provide a general overview of how extension mechanisms incorporate additional data elements that are not defined in the current schema of the IFC standard.

4.3.1 Property Set Definitions in IFC

The concept of property set definition is used to specialize and extend IFC entities without the requirement to define new subtypes. There are two types of property sets that can be assigned to IFC entities, namely, statically defined property sets (instances of IfcPreDefinedPropertySet) and dynamically extendable property sets (instances of IfcPropertySet).

The statically defined property sets are part of the IFC specification and are mainly used for enriching some architectural elements with more attributes.

"IfcPreDefinedPropertySet" is the abstract entity for all types of statistically defined property sets. The subtypes of this entity are IfcDoorLiningProperties, IfcDoorPanelProperties, IfcPermeableCoveringProperties, IfcReinforcement DefinitionProperties, IfcWindowLiningProperties, and IfcWindowPanel Properties. These statistically defined properties are used for providing additional information for their corresponding element, if such information is required. For instance, IfcDoorPanelProperties is applied to the "IfcDoor" entity and provides a parametric description of width, depth, position, shape, style, and operation of a door panel. Although the statically defined property sets are not relevant to urban land administration, they can be considered as good examples for defining new dynamic properties.

In the context of this book, the dynamically extendable property sets are relevant to the enrichment of the IFC standard for urban land administration. These property sets refer to those properties where the IFC standard simply provides a type of "meta model," to be further defined for a specific domain. This means there is no entity definition for this type of property set within the IFC schema. They can be instantiated by pairing a string value to the Name attribute of "IfcPropertySet." The convention for naming a property set is "Pset_xxx." "IfcProperty" references the "IfcProperty" entity via the "HasProperties" attribute to assemble and group a specific set of properties. "IfcProperty" is an abstract generalization for subtype entities defining specific kinds of properties. Commonly used subtype entities for defining properties are IfcPropertySingleValue, IfcPropertyEnumeratedValue, IfcPropertyBoundedValue, and IfcPropertyReferenceValue.

IfcPropertySingleValue specifies a property with a single value. It is the most commonly used type of property. IfcPropertyEnumeratedValue provides the capability to select a value from a predefined set of alternatives. The list of actual values is stored in the attribute EnumerationValues. IfcPropertyBoundedValue defines a type of property with values varying between an upper limit and lower limit which are specified by the UpperBoundValue and LowerBoundValue attributes, respectively.

IfcPropertyReferenceValue entity is used for referencing those entities defined within the static schema of the IFC standard. This entity has the PropertyReference attribute referring to a selection data type called IfcObjectReferenceSelect to choose the entity type. However, only entities which are defined within the resource layer of the IFC schema can be selected via this selection data type. The reason is that the IfcPropertyReferneceValue entity is defined within the resource layer of the IFC standard, and it cannot reference entities defined in the above layers. This constraint is compliant with the technical architecture of the IFC standard that an IFC entity must only reference another entity in its container layer or a lower layer.

Both statically defined and dynamically extendable property sets can be paired to their corresponding IFC entities via the "IfcRelDefinesByProperties" relationship. This relationship provides the capability to enable independence of the property set from its corresponding IFC entity. Therefore, the property set can be assigned to the entity whenever it is required to do so. In addition, the "IfcRelDefinesByProperties" relationship can be used to assign a single property set to more than one IFC entity. This implies that several IFC entities could use a particular property set simultaneously, if required. "IfcRelDefinesByProperties" has two attributes: RelatedObjects and the RelatingPropertyDefinition. The RelatedObjects attribute refers to the object(s) (concrete subtypes of IfcObjectDefinition) to which the property set applies. The RelatingPropertyDefinition attribute points to the entity defining property set, which is an instance of "IfcPropertySet" for dynamic properties.

4.3.2 User-Defined Values in IFC

The majority of IFC entities include a "USERDEFINED" enumeration or attribute to specify semantically a particular IFC data element which doesn't exist in the IFC schema. For example, the "IfcActorRole" entity includes the UserDefinedRole element for defining those actor roles that are not already considered in the IFC standard. To define the land surveyor as an actor in the IFC standard, the "LAND SURVEYOR" value, that is not already defined as the value of the UserDefinedRole attribute, is used.

4.4 Modeling Legal Interests in IFC

Each legal interest can be modeled in the IFC standard by considering two aspects: attributes and spatial structure. The attributes of a legal interest will be defined as dynamic property sets. The spatial structure of a legal interest can be composed of either a single element or multiple elements. Therefore, an explanation of how spatial or physical elements can be used for modeling each legal interest with individual part or multiple parts is provided. Multi-part legal interests must be defined by pairing the physical and spatial elements in non-hierarchical spatial structures. These structures in the IFC standard are specified by two types of zones: zone (IfcZone) and spatial zone (IfcSpatialZone).

A zone is defined as a group of internal spaces (IfcSpace) as well as already defined zones and spatial zones. The "IfcRelAssignsToGroup" objectified relationship handles the assignment of constituent parts, which are typically instances of "IfcSpace," to the whole structure of a zone. The grouping relationship can be used in a recursive way, which means one zone can be grouped under another zone. A zone does not have its own shape and representation.

A spatial zone is composed of invisible spatial elements (concrete subclasses of "IfcSpatialElement") in addition to the physical elements (concrete subclasses of "IfcElement") in a non-hierarchical method for a specified purpose (e.g. thermal or lighting activity of the building). Spatial zone can be considered as the suitable concept for managing legal interests associated with a combination of cognitive spaces and physical elements spanned throughout the building developments. Spatial zones are more overarching concepts than zones since in addition to internal spaces, external spaces and physical elements can have a direct relationship with the whole structure of a spatial zone.

The IFC standard maintains different types of spatial zones in buildings. As indicated in Figure 4.1, the PredefinedType attribute uses the IfcSpatialZoneTypeEnum enumeration data type to help identify the type of spatial zone. The "IfcSpatialZone" entity is related to its constituent elements via the "IfcRelReferencedInSpatialStructure" objectified relationship. This relationship makes use of the RelatedElements attribute to point to a set of elements (subclasses of "IfcProduct" entity) referenced in a spatial zone (see Figure 4.1). In other words, the "IfcRelReferencedInSpatialStructure" relationship entity is used to formulate what elements define a spatial zone for each legal interest.

In addition, the attributes of each legal interest can be defined as new property sets (instances of "IfcPropertySet"). These property sets can be paired to their corresponding legal interest via the "IfcRelDefinesByProperties" relationship (see Figure 4.1). Another important factor in assigning property sets is that they can also be attached to each element constituting a specific legal spatial zone. For instance, when a volumetric lot only consists of

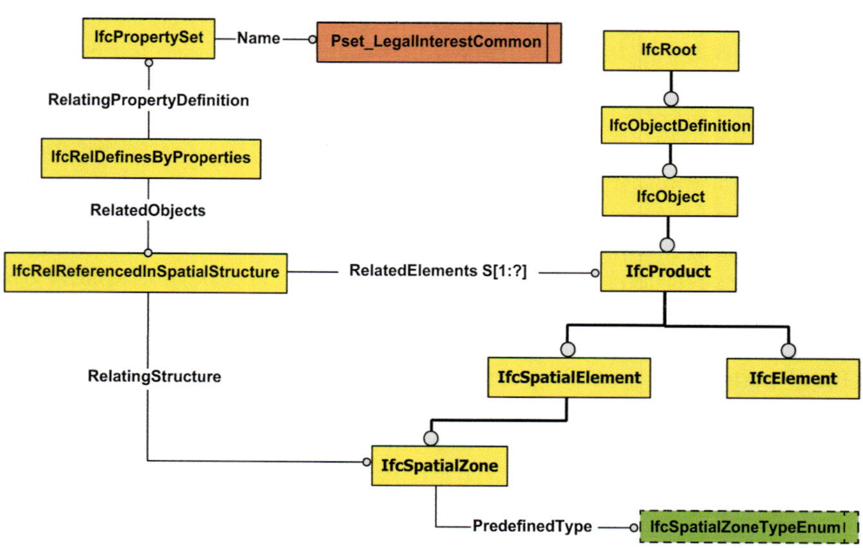

FIGURE 4.1
Spatial zones and assignment of property sets.

TABLE 4.1

Proposed Property Set Shared Among All Legal Interests

Property Set Name	**Pset_LegalInterestCommon**	
Property Name	Property Type	Data Type
LegalInterestName	IfcPropertySingleValue	IfcLabel
LegalInterestUnit	IfcPropertyEnumeratedValue	IfcLabel
LegalInterestState	IfcPropertyEnumeratedValue	IfcLabel
LegalInterestClass	IfcPropertyEnumeratedValue	IfcLabel
LegalInterestArea	IfcPropertySingleValue	IfcAreaMeasure
LandUse	IfcPropertyEnumeratedValue	IfcLabel

one space, both the "IfcSpace" and "IfcSpatialZone" entities can be used to model the volumetric lot. In this case, proposed property sets for volumetric lots can be assigned to either "IfcSpace" or "IfcSpatialZone" depending on which entity was used to model the volumetric lot. Therefore, a list of IFC entities, in addition to "IfcSpatialZone," to which each property set is applicable will be provided. Some properties (attributes) are shared among all legal interests. Table 4.1 provides the shared property set, which can be assigned to any legal interest. Depending on the type and spatial arrangement of a legal interest, this property set can be applied to a wide range of spatial and physical IFC entities such as IfcSpatialZone, IfcZone, IfcSpace, IfcSite, IfcExternalSpatialElement, IfcCivilElement, IfcGeographicElement, IfcBuildingElement and IfcDistributionElement.

4.4.1 Primary Legal Interests in IFC

In Section 2.4.1.1, four types of primary legal interests were identified in the context of Victorian jurisdiction: lot, common property, road, and reserve. We will now explain how each primary legal interest will be modeled inside the IFC standard.

4.4.1.1 Lot

For a lot associated with the vacant land or individually owned buildings, its legal interest is spatially defined by site elements showing the land surface of the lot. This type of lot is called "land lot" in this book. A single part land lot is defined by one instance of the IfcSite entity. To define a multi-part land lot inside the IFC standard, the concept of spatial zone should be considered. The spatial zone of a land lot references a set of instances of the "IfcSite" entity (see Figure 4.2a). The spatial zone can have its own placement and geometric representation to show the legal extent of a land lot. The land lot modeled inside the IFC standard comprises 2.5D parcels. This indicates that the boundary points include three physical coordinates. Table 4.2 provides the specific attributes (Pset_Lot) for any type of lot.

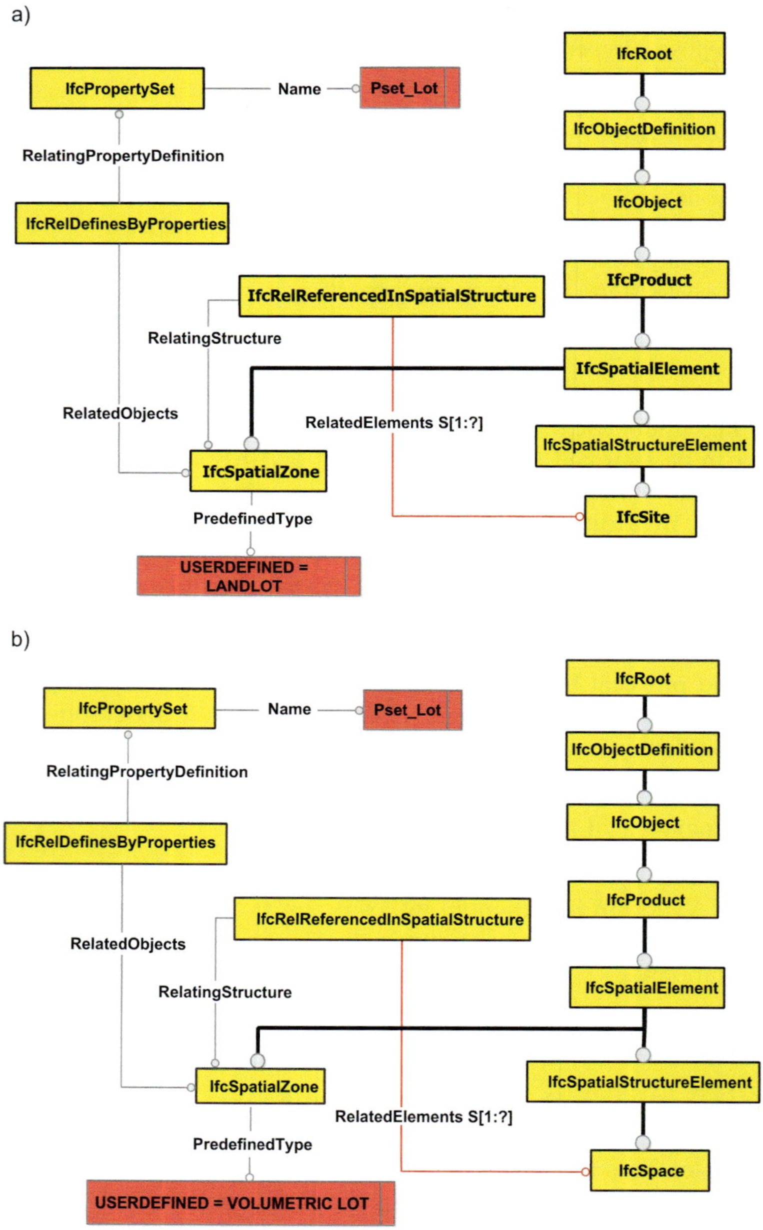

FIGURE 4.2
(a) Modeling a multi-part land lot in IFC; (b) Using the IfcSpatialZone entity to model a multi-part volumetric lot in IFC.

TABLE 4.2

Proposed Properties for Lot

Property Set Name	**Pset_Lot**	
Property Name	Property Type	Data Type
LotLiability	IfcPropertySingleValue	IfcPositiveRatioMeasure
LotEntitlement	IfcPropertySingleValue	IfcPositiveRatioMeasure

TABLE 4.3

Other Existing Properties of Lot

Property Name	**Corresponding IFC entity**	**Relevance to Lots**
GrossVolume	Qto_SpaceBaseQuantities which is applied to IfcSpace	This quantity can be used to determine the volume of a volumetric lot.
BuildingAddress	Attribute of IfcBuilding	This attribute can be used to provide the address of lots.

Moreover, Table 4.3 includes the other important attributes of the lot, which already exist in the IFC standard. To manage the attributes of a land lot, the properties brought in Table 4.2 and Table 4.3 can be applied to either the IfcSpatialZone or IfcSite entity.

Lots inside multi-owned buildings can be typically modeled by referencing internal volumetric spaces such as an apartment unit, storage area or parking space. This type of lot is called a "volumetric lot" in this book. Another form of volumetric lot is the individual ownership associated with an airspace. A single part volumetric lot defined inside the building can be modeled by referencing one instance of the IfcSpace entity. An individual external legal space (i.e. airspace) can be modeled by using IfcExternalSpatialElement. An example of a single part lot is the legal space of an office block without any parking space or storage area. A multi-part volumetric lot can be modeled using either the IfcZone or IfcSpatialZone entity. A zone or spatial zone of this legal interest is defined by grouping or referencing a set of instances of the IfcSpace entity. An example of a multi-part volumetric lot is a zone or spatial zone comprising the legal space of an apartment unit as well as the legal spaces of its accessory parts such as a car park and storage area. Figure 4.2b shows how a multi-part volumetric lot can be modeled within the IFC standard using the concept of spatial zones. For volumetric lots, properties brought in Table 4.2 and Table 4.3 can be applied to the IfcSpatialZone, IfcZone, or IfcSpace entity.

4.4.1.2 Common Property

Common properties normally have a complex spatial structure. It is not easy to generalize all the physical or spatial elements defining the spatial arrangement of common properties. This type of legal interest typically includes a combination of communal land parcels and/or legal spaces defined as the internal and external elements of building developments. Furthermore, a wide range of building elements (including boundary walls, ceilings, floors, windows, doors, and so on), and utility elements (such as pipes and electrical wiring) inside buildings are considered to be a part of common property.

Therefore, a single part common property can be modeled by an instance of the IfcSpace, IfcExternalSpatialElement, or IfcSite entity. Another rare case for defining a single part common property can be based on an individual physical structure. In this case, a common property is defined by an instance of a concrete subtype of the IfcElement, such as IfcBuildingElement and IfcDistributionElement.

In most cases, common properties are spatially composed of multiple parts in urban land administration practices. The spatial arrangement of multi-part common properties can be defined as IfcZone or IfcSpatialZone. If a common property comprises only internal spaces, IfcZone can be used to model its spatial structure. Although it is possible to group instances of IfcSpatialZone inside the IfcZone, it is suggested that the IfcSpatialZone entity is used for modeling those common properties which comprise external spaces, physical structures, and land parcels. This is due to the fact that using the IfcSpatialZone entity provides the ability to define a direct relationship between the whole spatial structure of the common property and its constituent spatial or physical elements (see Figure 4.3).

In addition, Table 4.4 provides the specific attributes (Pset_CommonProperty) that can be applied to the IFC entities defining the common property. These entities can be IfcSpatialZone, IfcZone, IfcSite, IfcSpace, IfcExternalSpatialElement, and concrete subtypes of IfcElement.

4.4.1.3 Road

The physical representation of roads in the IFC standard are defined by the IfcGeographicElement and IfcCivilElement entities. IfcGeographicElement is used for modeling existing roads, while IfcCivilElement is used for modeling new roads. However, in urban land administration practices, the legal extent of roads is considered as land parcels. In the context of 3D digital urban land administration, it is better to consider only the legal representation of roads, since the physical objects defining a road are spatially contained within the legal extent.

Therefore, a single part road can be modeled by an instance of the IfcSite entity. A multi-part road can be modeled by the IfcSpatialZone entity. A road spatial zone should reference a set of instances of the IfcSite entity (see Figure 4.4a). The attributes of a road parcel are only the shared ones mentioned in Table 4.1. If the legal interest is a road parcel, these attributes can be applied to either the IfcSpatialZone or IfcSite entity.

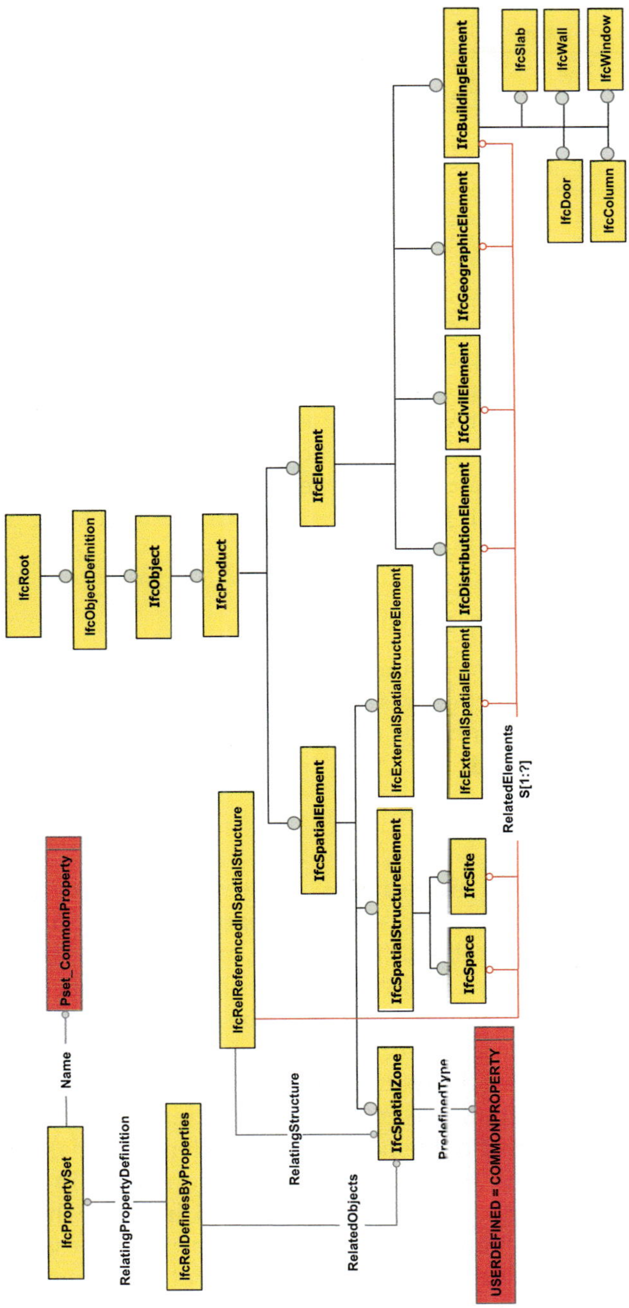

FIGURE 4.3
Modeling a multi-part common property as a spatial zone in IFC.

TABLE 4.4

Proposed Property Set for Common Property

Property Set Name	Pset_CommonProperty	
Property Name	Property Type	Data Type
CommonPropertyType	IfcPropertyEnumeratedValue	IfcLabel
LandAffected	IfcPropertySingleValue	IfcText
LimitationDetails	IfcPropertySingleValue	IfcText

4.4.1.4 Reserve

Reserves are legally defined as land parcels. Therefore, an instance of the IfcSite entity can be used to model the legal extent of a single part reserve. If a reserve includes multiple parts, then the IfcSpatialZone entity can be used to model the whole spatial structure of the reserve by referencing a set of instances of the IfcSite entity (see Figure 4.4b). The spatial zone defined for a reserve can have its own placement and geometric representation to show its legal extent. The attributes of reserves are the only shared attributes illustrated in Table 4.1. For reserves, they can be applied to either the IfcSpatialZone or IfcSite entity.

4.4.2 Secondary Legal Interests in IFC

In Section 2.4.1.2, four types of primary legal interests were identified in the context of Victorian jurisdiction: easement, restriction, depth limitation, and airspace. We will now explain how each primary legal interest will be modeled inside the IFC standard.

4.4.2.1 Easement

The spatial arrangement of an easement can be defined based on either an individual object or multiple objects. A single part easement can be spatially defined as either a land parcel, a utility element, or a legal space surrounding the utility element. Therefore, the IfcSite, IfcDistributionElement (and its subtypes), or IfcExternalSpatialElement entity can be used to define a single part easement. For multi-part easements, a wide range of spatial and physical elements can be used to define the spatial extent of the easement. Therefore, it is suggested that the IfcSpatialZone entity, which references a set of instances of the IfcSite, IfcDistributionElement, and IfcExternalSpatialElement, is used for modeling a multi-part easement (see Figure 4.5a). In addition to the property set illustrated in Table 4.1, Table 4.5 includes the attributes (Pset_EasementOrRestriction) which are shared between easements and restrictions. The specific attributes (Pset_Easement) of the easement are also provided in Table 4.6. For easements, both the "Pset_RestrictionCommon" and "Pset_Easement" property sets can be applied to the IfcSpatialZone, IfcSite, IfcDistributionElement, or IfcExternalSpatialElement entity.

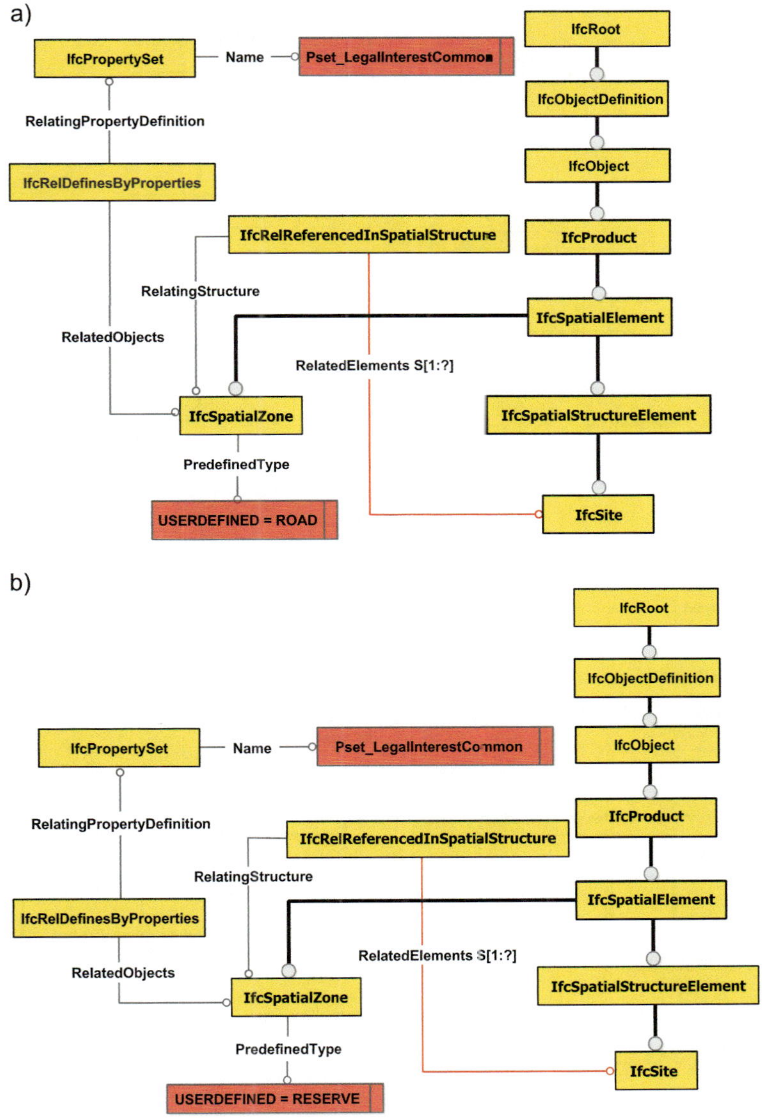

FIGURE 4.4
(a) Spatial zone defined for modeling a multi-part road in IFC; (b) The spatial zone defining a multi-part reserve in IFC.

4.4.2.2 Restriction

A restriction can be applied to either a single spatial object or multiple spatial objects. The spatial extent of a restriction can be defined by site elements showing the land surface for the restricted area. In addition, the restriction can also be applied to external spaces around buildings, such as building

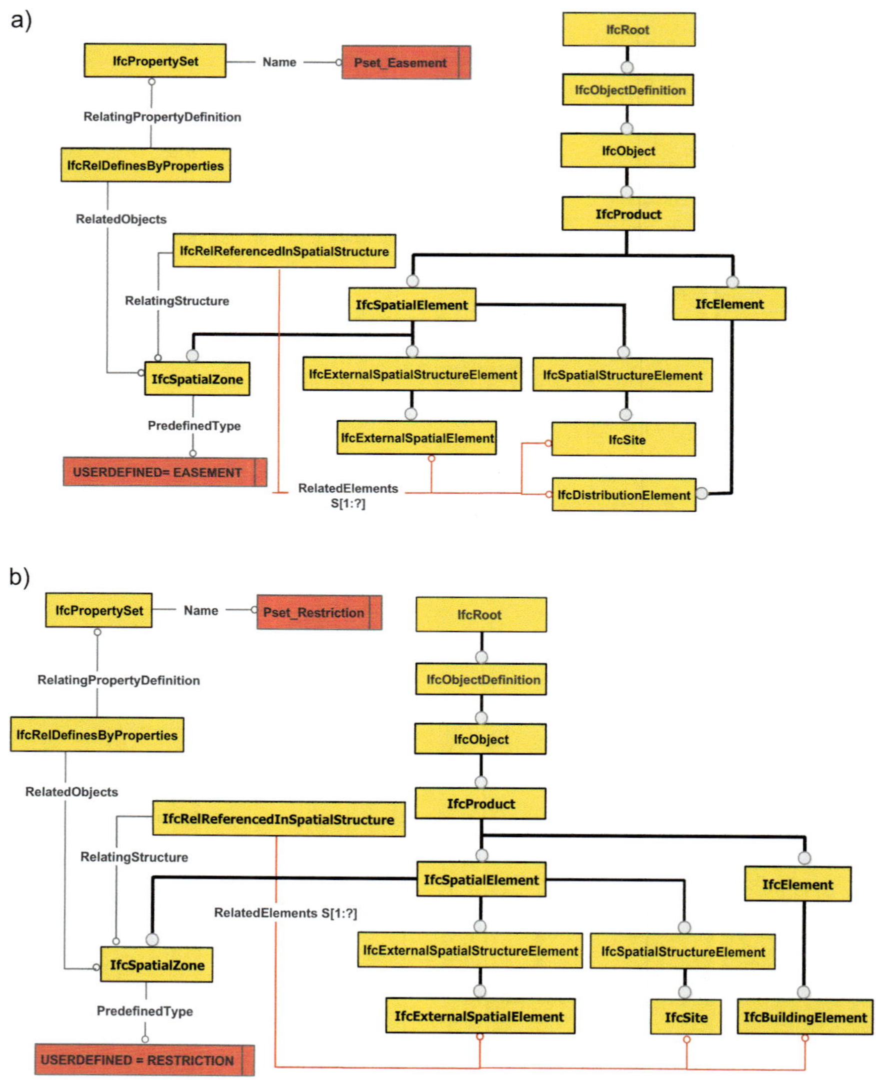

FIGURE 4.5
(a) The spatial zone defining a multi-part easement in IFC; (b) The spatial zone defining a multi-part restriction in IFC.

TABLE 4.5
Property Set Shared Between Easement and Restriction

Property Set Name	**Pset_EasementOrRestriction**	
Name	Property Type	Data Type
LandBenefited	IfcPropertySingleValue	IfcText
LandBurdened	IfcPropertySingleValue	IfcText

TABLE 4.6

The Property Set Specifically Applied to Easements

Property Set Name	Pset_Easement	
Name	Property Type	Data Type
EasementWidth	IfcPropertySingleValue	IfcLengthMeasure
EasementOrigin	IfcPropertySingleValue	IfcLable
EasementPurpose	IfcPropertySingleValue	IfcLable

TABLE 4.7

Specific Property Set for Restriction

Property Set Name	Pset_Restriction	
Name	Property Type	Data Type
RestrictionDescription	IfcPropertySingleValue	IfcText
RestrictionExpiryDate	IfcPropertySingleValue	IfcDate

envelopes, and materials of building elements, such as colors and setbacks of walls. Therefore, a single part restriction can be defined by an instance of the IfcSite, or IfcExternalSpatialElement, or IfcBuildingElement entity (see Figure 4.5b). For multi-part restrictions, the IfcSpatialZone entity should be adopted. This spatial zone is defined by referencing a set of instances of the IfcSite, IfcExternalSpatialElement, and IfcBuildingElement entities. In addition to the property sets brought in Table 4.1 and Table 4.5, the attributes (Pset_Restriction) provided in Table 4.7 should also be used to manage information describing restrictions. For restrictions, these attributes can be applied to an instance of the IfcSpatialZone, IfcSite, IfcExternalSpatialElement, and IfcBuildingElement entities.

4.4.2.3 Depth Limitation and Airspace

Depth limitations are spatially defined as legal spaces located below the urban developments. An individual depth limitation can be modeled by an instance of the IfcExternalSpatialElement. If a depth limitation has multiple parts within an urban development, the IfcSpatialZone entity can be used to define a spatial zone containing multiple instances of the IfcExternalSpatialElement entity (see Figure 4.6a). Modeling airspaces inside the IFC standard follows a similar approach. The spatial extent of airspaces can be defined by either an individual legal space or multiple legal spaces surrounding urban developments. Therefore, the IfcExternalSpatialElement and IfcSpatialZone entities are the suitable entities for modeling airspaces (see Figure 4.6b). The specific attributes (Pset_DepthLimitationOrAirspace) of a depth limitation or airspace are illustrated in Table 4.8. These attributes can be applied to the IfcSpatialZone and IfcExternalSpatialElement entities (see Figures 4.9 and 4.10).

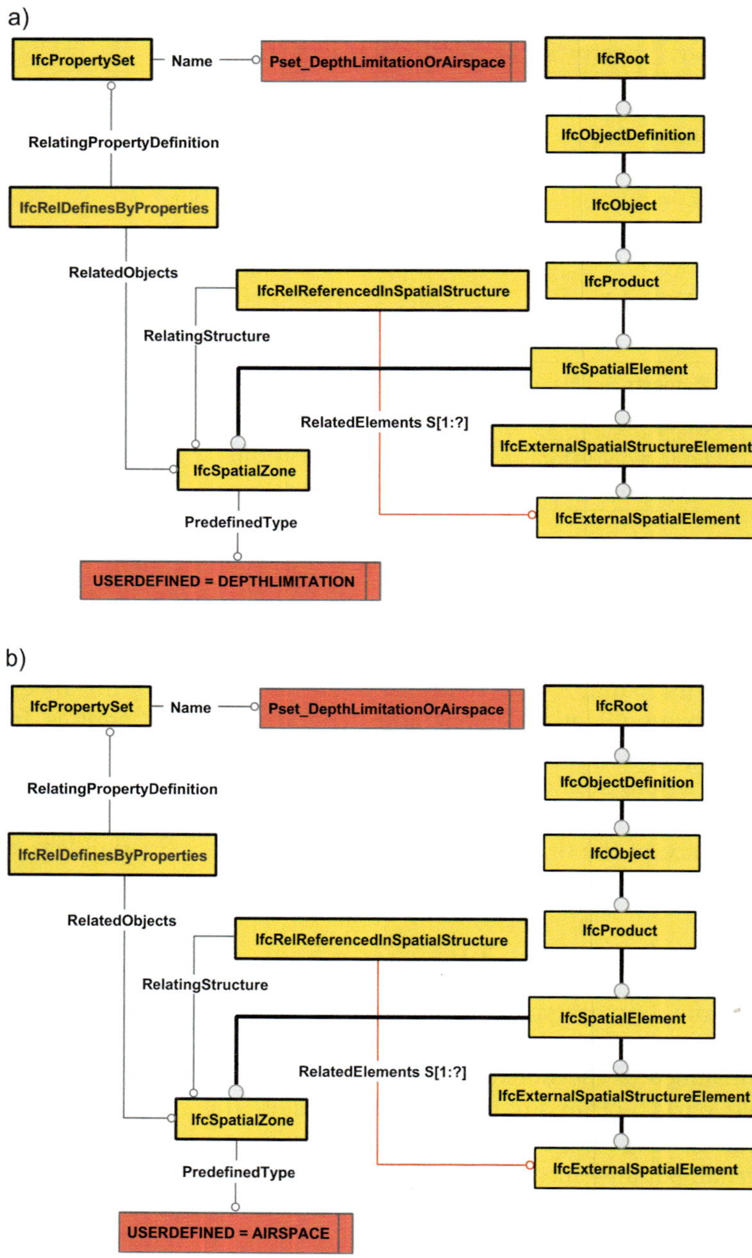

FIGURE 4.6
(a) Modeling a multi-part depth limitation in IFC; (b) Modeling a multi-part airspace in IFC.

TABLE 4.8

Specific Property Set for Depth Limitation and Airspace

Property Set Name	**Pset_DepthLimitationOrAirspace**	
Property Name	Property Type	Data Type
Length	IfcPropertySingleValue	IfcLengthMeasure
LandApplied	IfcPropertySingleValue	IfcText
UpperLimit	IfcPropertySingleValue	IfcLengthMeasure
LowerLimit	IfcPropertySingleValue	IfcLengthMeasure

4.5 Modeling Legal Boundaries in IFC

The IFC standard comprises a wide range of semantic, geometric, and topologic entities used for modeling both line and surface boundaries. Line boundaries are used to define the spatial limits of 2D or 2.5D land parcels, while the spatial dimensions of 3D legal spaces are defined by surface boundaries. Within the IFC schema, the IfcRelSpaceBoundary entity plays an important role in modeling both the geometry and semantics of surface boundaries related to the various legal interests. In the following subsections, entities used for modeling the geometry of legal boundaries will first be explained in detail. Subsequently, a detailed description of semantic entities required for distinguishing different types of general and fixed boundaries within the IFC standard will be provided.

4.5.1 Modeling the Geometry and Topology of Legal Boundaries in IFC (4)

Within the IFC standard, IfcConnectionGeometry is the general entity used to define the geometry and topology of physical connections between two spatial objects. This entity has four subtype entities to model various kinds of physical connections. These subtype entities are:

- IfcConnectionPointGeometry: for modeling physical connections at a geometric point or a vertex associated with point coordinates
- IfcConnectionCurveGeometry: for modeling physical connections at a curve or an edge associated with a curve geometry
- IfcConnectionSurfaceGeometry: for modeling physical connections at a surface or a face associated with a surface geometry
- IfcConnectionVolumeGeometry: for modeling physical connections at a volume or a shell associated with a volume geometry

Among these physical connections, a physical connection at a curve or an edge can be used for modeling line-based legal boundaries

(IfcConnectionCurveGeometry), while a physical connection at a surface or a face is considered for modeling surface-based legal boundaries (IfcC onnectionSurfaceGeometry).

4.5.1.1 Line-Based Legal Boundaries

For legal interests associated with land parcels, such as reserves and roads, the geometry and topology of legal boundaries should be modeled as curves and edges. The IfcConnectionCurveGeometry entity is used to define boundaries between land parcels with 2D spatial extent. As illustrated in Figure 4.7a, this entity includes two attributes: CurveOnRelatingElement and CurveOnRelatedElement. The CurveOnRelatingElement attribute is mandatory and defines the line boundary at which two land parcels are connected. In addition, the optional "CurveOnRelatedElement" relationship provides the geometry and placement of the same curve within the local coordinate system of the related land parcel. The spatial location of the boundary is defined with respect to the local coordinate system of its relating land parcel. The CurveOnRelatingElement attribute refers to the IfcCurveOrEdgeCurve selection data type. This data type provides the ability to specify whether the boundary is a bounded curve (IfcBoundedCurve) or an edge with an associated curve (IfcEdgeCurve). Here, there are two selections: purely geometric modeling of the boundary line by IfcBoundCurve and topologic-geometric modeling of the boundary line by IfcEdgeCurve.

Thegeometricmodelingoftheboundarylineisdefinedbythe IfcBoundedCurve entity which has four subtypes to model various types of curves: spline curves (IfcBSplineCurve), continuous curves (IfcCompositeCurve) composed of curve segments, bounded curves (IfcPolyline) with linear segments defined by a list of Cartesian points, and trimmed curves (IfcTrimmedCurve). Although it is sometimes possible to delineate curved legal boundaries, linear legal boundaries are mainly adopted in land administration practices. Therefore, the IfcPolyline entity is mainly suggested for defining the geometry of legal boundary lines inside the IFC standard. The IfcPolyLine entity also includes a condition which specifies that all the points defining the boundary line have the same spatial dimensionality. This condition ensures that all the points defining boundary lines are either 2D points or 3D points.

In addition to the geometry, topology of the legal boundary line should be considered. Therefore, IfcEdgeCurve could be a good choice for the spatial modeling of line-based legal boundaries. An instance of IfcEdgeCurve is defined by two vertices which are connected topologically and include the geometric representation of the boundary line. The IfcEdgeCurve entity includes two attributes:

- EdgeGeometry: This attribute refers to the curve defining the shape and geometry of the edge. Multiple edges can refer to the various parts of the same curve.

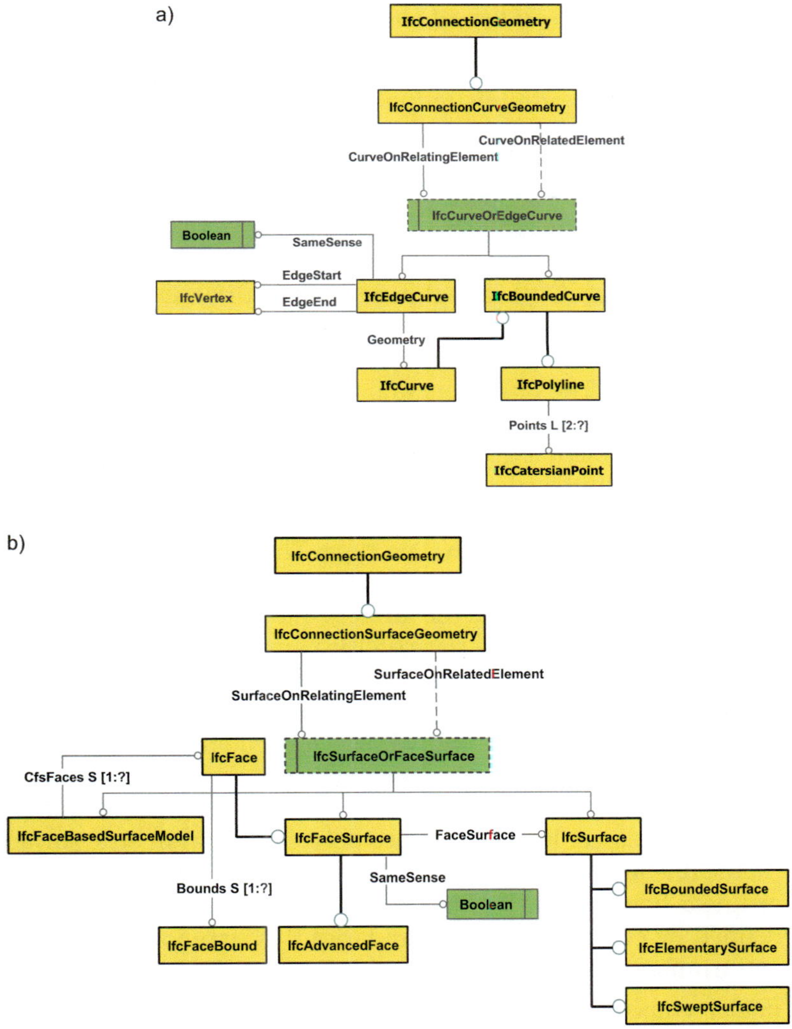

FIGURE 4.7
(a) Relevant entities for modeling the geometry and topology of line-based legal boundaries in IFC; (b) Relevant entities for modeling the geometry and topology of surface-based legal boundaries in IFC.

- SameSense: This Boolean attribute specifies if the sense of the edge is the same as the sense of the curve providing the geometry. The sense of an edge is defined from the start vertex towards the end vertex. The sense of the curve is typically defined in the direction of increasing parameter. For the legal boundary lines, the geometric direction of the curve should be the same as the topological direction of the edge. Therefore, the value of this attribute should be "TRUE" for all legal boundary lines.

4.5.1.2 Surface-Based Legal Boundaries

Internal and external volumetric spaces are often considered for defining legal interests in urban land administration practices. The ConnectionGeometry attribute of the IfcRelSpaceBoundary entity is used to define the geometry and topology of legal boundary surfaces between two 3D legal spaces. This attribute refers to the IfcConnectionGeometry entity; however, its subtype entity IfcConnectionSurfaceGeometry is used for modeling surface-based legal boundaries (see Figure 4.7b). IfcConnectionSurfaceGeometry includes two attributes: SurfaceOnRelatingElement and SurfaceOnRelatedElement. The SurfaceOnRelatingElement attribute is mandatory and defines the surface boundary at which two legal spaces or a legal space and a physical element are connected. The placement and geometry of the boundary surface are described relative to its relating legal space. In addition, the optional CurveOnRelatedElement attribute provides the geometry and placement of the same surface within the local coordinate system of the related legal space or physical element. The SurfaceOnRelatingElement attribute refers to the IfcSurfaceOrFaceSurface selection data type. This data type provides three alternatives for defining surface-based legal boundaries: purely geometric modeling of a boundary surface (IfcSurface), integrated geometric and topologic modeling of a boundary surface (IfcFaceSurface), and a connected set of faces defining a surface model (IfcFaceBasedSurfaceModel).

IfcSurface provides the most generalized representation of a 2D surface located within a 3D space. This entity is specialized into three subtypes:

- IfcBoundedSurface: This entity refers to a category of surfaces with a finite area. Its subtypes are polynomial parametric surfaces (IfcBSplineSurface), parametric planar surfaces defined by curved or linear boundaries (IfcCurveBoundedPlane), parametric surfaces defined by curved or linear boundaries (IfcCurveBoundedSurface), and trimmed surfaces (IfcRectangularTrimmedSurface).
- IfcElementarySurface: This entity is an abstract entity when defining two types of simple analytical surfaces: cylindrical surfaces unbounded in the vertical dimension (IfcCylindricalSurface) and planar surfaces unbounded in the vertical dimension (IfcPlane).
- IfcSweptSurface: Swept surfaces are defined by sweeping a curve or line. There are two types of swept surfaces: one created by sweeping a curve or a line alongside a particular direction defined by a vector (IfcSurfaceOfLinearExtrusion) and a surface derived from rotating a curve or a line alongside an axis (IfcSurfaceOfRevolution).

Among the above-mentioned surface types, bounded surfaces can mainly be used to represent the geometry of 3D legal spaces defined in urban land administration practices. This is due to the fact that the spatial extent of these surfaces should be finite and their bounding lines or curves should be defined.

However, other surface types can be also used in some cases. For instance, when the ownership extent of a land parcel is defined from the earth center to infinity, elementary surfaces can be a good choice to model these types of infinite legal spaces. Swept surfaces can be useful in the delineation stage of legal boundaries; however, they should be converted to bounded surfaces in the stage of representation and registration of 3D legal spaces.

In addition to the geometry of the legal boundary surface, the topology should also be considered. Therefore, IfcFaceSurface could be a good choice for spatial modeling of surface-based legal boundaries. A face surface (IfcFaceSurface) is a particular type of face (IfcFace) which can be used to model both the geometrical and topological aspects of surface-based legal boundaries. For this purpose, the IfcFaceSurface entity includes three attributes:

- Bounds: This attribute is inherited from IfcFace, which refers to a loop (IfcFaceBound) defining the boundaries of the face.
- FaceSurface: This refers to the geometric surface defining the internal shape of the face. The realm of the face is composed of this surface and its bounding loop.
- SameSense: This Boolean attribute specifies if the normal direction of the face is the same as the normal direction of the surface providing the geometry. For legal boundary surfaces, the geometric normal direction of the surface should be the same as the topological normal direction of the edge. Therefore, the value of this attribute should be "TRUE" for all legal boundary surfaces.

For modeling face surfaces with complex shapes, the IfcAdvancedFace entity can be used. This entity is a specialized subtype of the IfcFaceSurface entity. An advanced face is defined based on some particular requirements to define its geometry and topology. These requirements are:

- The geometry of the face surface must be limited to elementary surfaces (IfcElementarySurface), swept surfaces (IfcSweptSurface), or a polynomial parametric surface (IfcBSplineSurface).
- The outer bound of the face shall be defined by only an instance of the IfcFaceOuterBound entity.
- The face is restricted to be bound by IfcEdgeLoop or IfcVertexLoop.
- Every edge defining the face must be associated with a curve geometry.
- The geometry of the edge must be restricted to unbounded lines (IfcLine), parameterized planar curves (IfcConic), bounded curves with only linear segments (IfcPolyLine), and spline curves (IfcBSplineCurve).

IfcSurface and IfcFaceSurface are used for modeling an individual surface or face surface associated with a legal boundary. However, it is also possible to model a connected set of faces (IfcConnectedFaceSet) associated with a face-based surface model which is defined by an instance of the IfcFaceBasedSurfaceModel entity. The faces of a face-based surface model are connected arcwise, and they can only connect with each other at edges or vertices. The IfcConnectedFaceSet includes the CfsFaces attribute which refers to a set of instances of the IfcFace entity.

4.5.2 Semantic Modeling of Legal Boundaries in IFC

In Section 2.4.1.3, various boundary types used in urban land administration practices were identified. This section explains how the semantics of each boundary type can be modeled within the IFC standard. It will describe how boundary types which are not semantically considered inside the IFC standard can be semantically stored in the IFC standard, in line with the logic of its data model.

4.5.2.1 General Boundaries in IFC

General boundaries are typically defined by referencing physically existent objects. To define general boundaries inside the IFC standard, the IfcRelSpace Boundary uses the RelatedBuildingElement attribute which references those IFC entities defining physical elements (see Figure 4.8). Additionally, the value of the IfcPhysicalOrVirtualEnum attribute is required to be set to PHYSICAL since general boundaries reference physically existent spatial objects. The conditional CorrectPhysOrVirt rule, which is defined inside the IfcRel SpaceBoundary entity, ensures that if the value of the IfcPhysicalOr VirtualEnum attribute is PHYSICAL, then the RelatedBuildingElement attribute does not refer to the IfcVirtualElement entity.

To define the semantics of building boundaries, the IfcRelatedBuilding Element attribute should refer to the IfcBuildingElement entity which generalizes building structures used for delineating the building boundary. Those building structures mainly used for defining building boundaries include walls (IfcWall), ceilings (IfcCeiling), roof (IfcRoof), slabs (IfcSlab), doors (IfcDoor), windows (IfcWindow), and similar architectural components. To semantically distinguish the location of a building boundary, the InternalOrExternalBoundary attribute refers to the IfcInternalOrExternalEnum enumeration. In the current version of the IFC standard, this enumeration includes the following values:

- INTERNAL: The boundary surface references a building structure which is connected to an internal legal space on the other side.
- EXTERNAL: The boundary surface references a building structure which is connected to an external legal space on the other side.

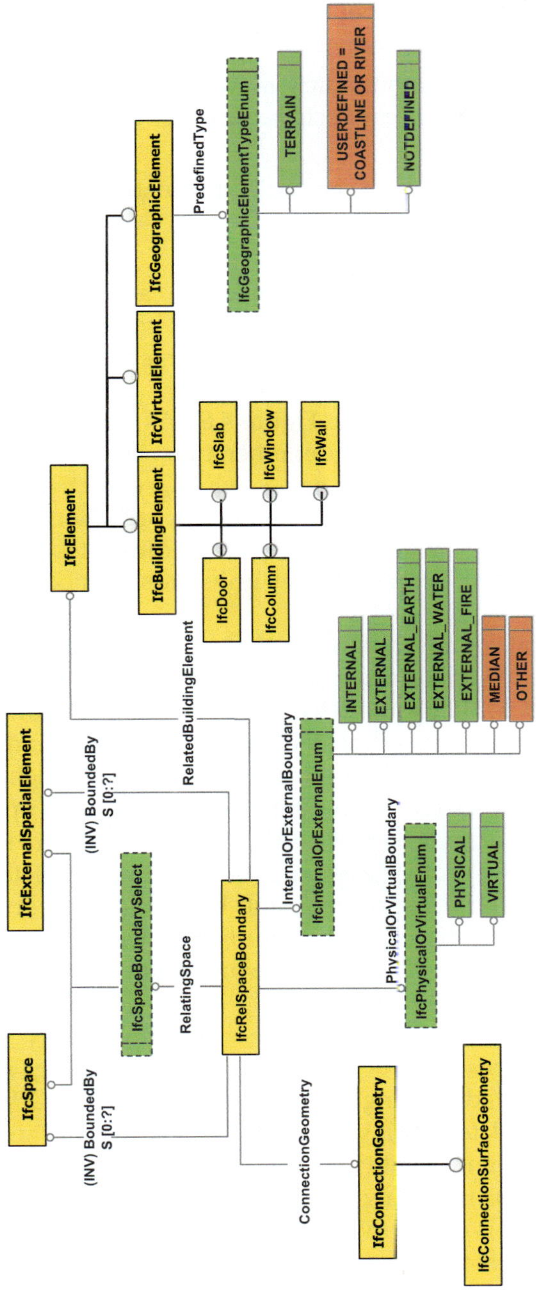

FIGURE 4.8
Semantic entities for modeling legal boundaries in the IFC standard.

- EXTERNAL_EARTH: The boundary surface references a physical or virtual element which is connected to the terrain on the other side.
- EXTERNAL_WATER: The boundary surface references a physical or virtual element which is connected to the water on the other side.
- EXTERNAL_FIRE: The boundary surface references a physical or virtual element which is connected to another building on the other side.
- NOTDEFINED: There is no semantic information provided for the boundary surface.

The interior and exterior building boundaries can be semantically differentiated from each other by specifying their corresponding enumerator (INTERNAL and EXTERNAL) from this enumeration. At present there is no enumerator defined for median and "other" building boundaries to differentiate them semantically (see Figure 4.8). As a result, it is suggested that the IfcInternalOrExternalEnum enumeration should be enriched with "MEDIAN" and "USERDEFINED" values which, respectively, provide the semantics for median and "other" boundary types referencing building structures. Figure 4.9a provides an example that illustrates how an interior wall boundary between two apartment units can be modeled using IFC entities.

While no physical manifestation exists for projected boundaries, these boundaries can be modeled in two ways. In the first approach, projected boundaries can be similarly modeled by referencing building elements, as projected boundaries are delineated through the extrusion of building boundaries. As an example, the boundary surface referencing a wall in a balcony area can be extended to delineate its projection up to the ceiling. Therefore, there is no apparent need to define a specific IFC entity for projected boundaries. Another approach to modeling projected boundaries is based on making a reference to the IfcVirtualElement entity since these boundaries themselves do not refer to any physical structure. In this method, the value of the IfcPhysicalOrVirtualEnum attribute must be set to VIRTUAL.

The last type of general boundaries are ambulatory boundaries. These boundary types can be modeled by considering specific values for the IfcRelatedBuildingElement and IfcPhysicalOrVirtualEnum attributes. The IfcRelateBuildingElement attribute should reference the IfcGeographicElemententity entity (see Figure 4.8). IfcGeographicElement is the semantic entity for modeling geographic features such as roads, trees, lakes, and rivers. In this entity, the "PredefinedType" attribute describes the type of geographic element. When modeling rivers and coastlines, this attribute should be set to "RIVER" and "COASTLINE" respectively. Additionally, the value of the IfcPhysicalOrVirtualEnum attribute should be EXTERNAL_WATER since ambulatory boundaries are defined based on waterbodies.

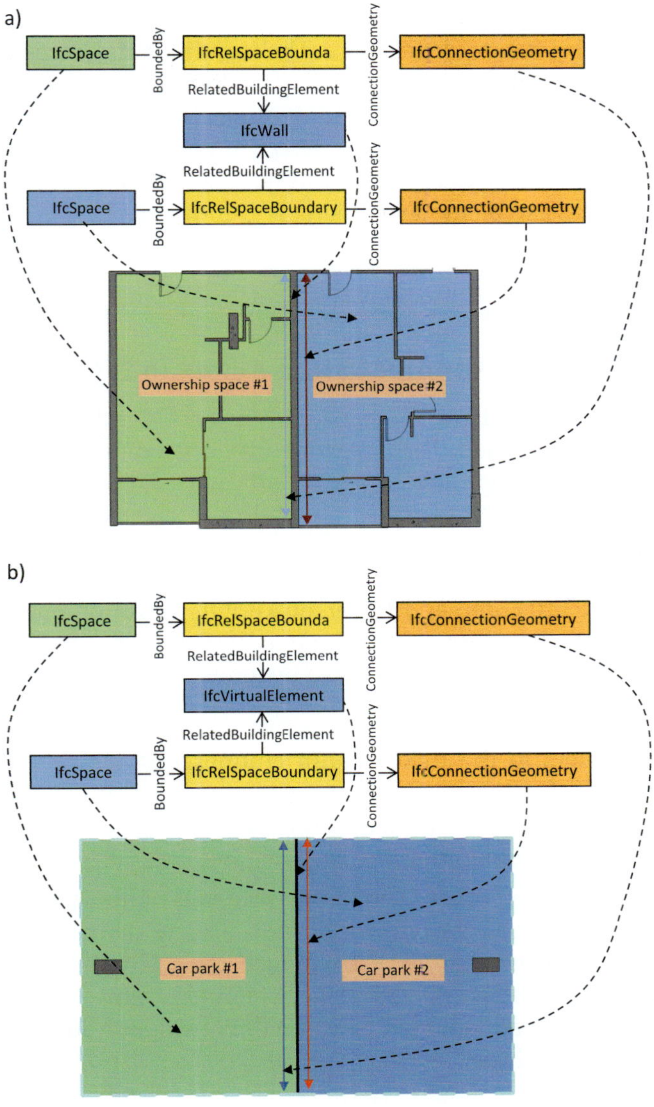

FIGURE 4.9
(a) Mapping an interior wall boundary within the IFC standard; (b) Mapping a fixed boundary within the IFC standard.

4.5.2.2 Fixed Boundaries in IFC

Fixed boundaries are defined based on surveying measurements and therefore do not reference a tangible spatial object in the real world. To semantically model fixed boundaries, the IfcRelSpaceBoundary entity should reference the IfcVirtualElement entity through the RelatedBuildingElement attribute (see Figure 4.8). In addition, the value of the IfcPhysicalOrVirtualEnum attribute must be set to VIRTUAL. The conditional CorrectPhysOrVirt rule, which is defined inside the IfcRelSpaceBoundary entity, ensures that if the value of the IfcPhysicalOrVirtualEnum attribute is VIRTUAL, then the RelatedBuildingElement attribute refers to the IfcVirtualElement entity. Figure 4.9b provides an example to illustrate how a fixed boundary between two car park areas can be modeled using IFC entities.

4.6 Modeling Land Administration Actors in IFC

To model a land administration actor within the IFC standard, the role of the actor should be defined. The IfcActor entity is used for defining all types of actors involved during the lifecycle of buildings. This entity includes the TheActor attribute, which provides the ability to choose from three types of actors: a person (IfcPerson), an organization (IfcOrganization), and a person acting on behalf of an organization (IfcPersonAndOrganization). All the IfcPerson, IfcOrganization, and IfcPersonAndOrganization entities include the Roles attribute which references the IfcActorRole entity to define the role performed by an actor. The IfcActorRole includes three important attributes:

- Role: This attribute provides the name of the role performed by an actor. It refers to the IfcRoleEnum enumeration which includes a wide range of values for various actor roles. Among land administration actors, the OWNER actor role has already been defined as a value for this enumeration.
- UserDefinedRole: This attribute provides the capability to define those actor roles which are not already defined within the IfcRoleEnum enumeration. For instance, the CITY COUNCIL actor role can be defined by this attribute. It should be noted that if a user-defined role is defined, the USERDEFINED value should be chosen for the Role attribute.
- Description: This attribute provides further information about the nature of the actor role.

Once the actors and their roles have been defined, they should be assigned to the objects for which they are responsible. The IfcRelAssignsToActor objectified relationship is used to handle the assignment of actors to the IFC objects. This assignment can be done using three attributes:

1. RelatingActor: It references the IfcActor entity which provides information describing the actor itself.
2. ActingRole: The role that the actor plays within the context of the assignment.

3. RelatedObjects: It refers to those objects, which are legal interests in this context, for which the actor is responsible.

Figure 4.10a illustrates an example of assigning a legal interest to a land administration actor. In this assignment, the common property is assigned to the Owners Corporation Manager with the acting role of "manage." Likewise, all other actors can be assigned to their corresponding legal interest using a similar approach. Table 4.9 provides a list of actor roles with their corresponding legal interests.

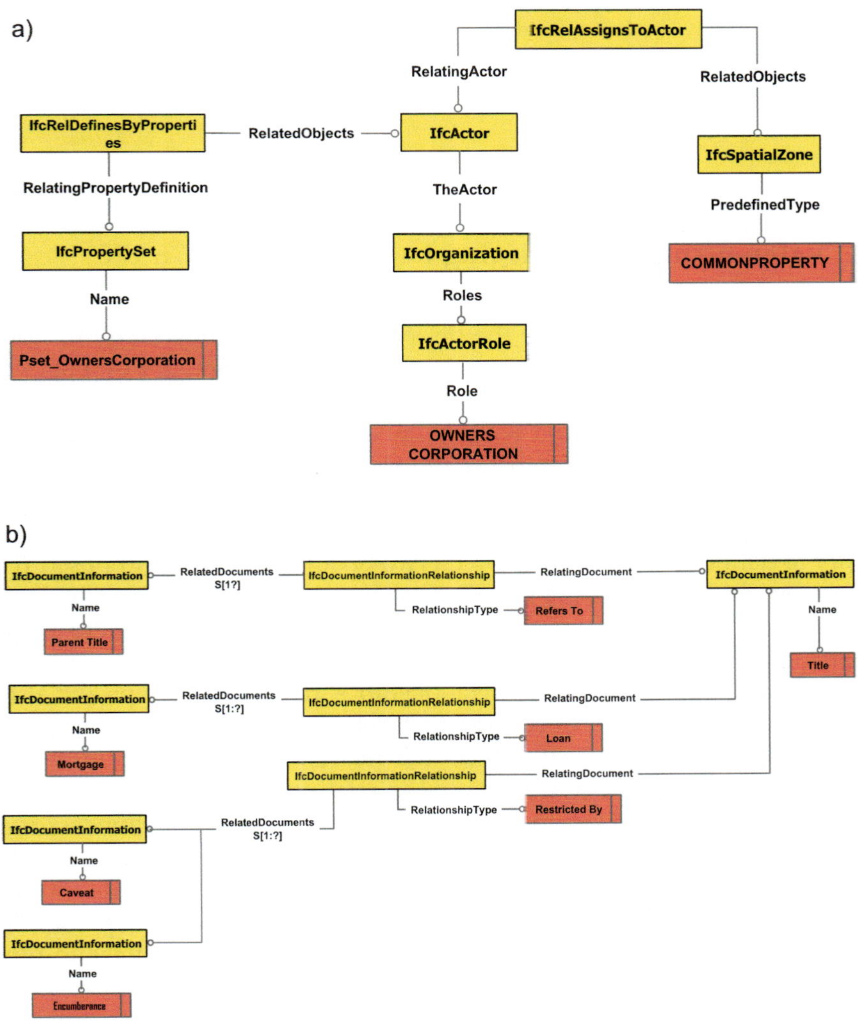

FIGURE 4.10
(a) Assigning Owners Corporation to common property; (b) Modeling legal documents used in the registration phase within the IFC standard.

TABLE 4.9

Actor Roles and Legal Interests

Actor Role	Actor Type	Acting Role	Legal Interest
Land surveyor	IfcPerson	Delineate	All legal interests
Owner	IfcPerson, IfcOrganization	Own	Volumetric Lot, Land Lot
Owners Corporation Manager	IfcPersonAndOrganization	Manage	Common Property
Land registry	IfcOrganization	Register	All legal interests
City Council	IfcOrganization	Certify	All legal interests
City Council	IfcOrganization	Own	Reserve
Referral Authority	IfcOrganization	Control, Own	Easement
Developer	IfcPerson, IfcOrganization	Develop	Land Lot

4.7 Referencing and Managing Legal Documents in IFC

When referencing and managing information relating to legal documents, it is a requirement to instantiate the IfcDocumentInformation entity for each legal document. This entity currently provides a rich amount of metadata for documents (see Table 4.10). The other required attributes for legal documents, such as Volume and Folio Number, can be provided through the Description attribute or referencing the URL for the location of legal documents. Table 4.11 includes these extra attributes for each legal document.

It is also possible to instantiate the IfcDocumentInformationRelationship entity to define relationships between two documents. In the proposed IFC extension, this entity is used to define relationships between the title document and other legal documents, such as the mortgage and caveat, created in the registration phase of the building subdivision process (see Figure 4.10b).

The IfcDocumentInformationRelationship entity has three specific attributes to relate two legal documents: (1) RelatingDocument: This attribute refers to the document that acts as the referencing document in a relationship. In the context of this book, the relating document is the "Title" document. (2) RelatedDocuments: This attribute refers to documents that act as the referenced documents in a relationship. The related legal documents here are "Parent Title," "Mortgage," "Caveat," and "Encumbrance." (3) RelationshipType: The value of this attribute describes the type of relationship between two legal documents. For example, the relationship type between "Title" and "Parent Title" is "Refers to," while "Restricted By" is the relationship type between "Title" and documents imposing restrictions on the tittle such as "Caveat" and "Encumbrance" (see Figure 4.10b).

Among the legal documents mentioned in Table 4.11, the title document is directly associated with primary legal interests, as well as their interest holders

TABLE 4.10

Current Attributes of the IfcDocumentInformation Entity (Extracted from ISO16739 2013)

Attribute	Description
Name	Document name
Description	Description of the document and its content.
Location	Resource identifier or locator provided as URI, URN, or URL, of the document information for online references.
Purpose	Purpose of the document.
IntendedUse	Intended use for this document.
Scope	Scope of this document.
Revision	Document revision designation.
DocumentOwner	Information about the person and/or organization acknowledged as the "owner" of this document. In some contexts, the document owner determines who has access to or editing right to the document.
Editors	The individuals and/or organizations who have created this document or contributed to it.
CreationTime	Date and time stamp when the document was originally created.
LastRevisionTime	Date and time stamp when this document version was created.
ElectronicFormat	Describes the media type used in various internet protocols, also referred to as "Content-type," or "MIME-type (Multipurpose Internet Mail Extension)," of the document being referenced.
ValidFrom	Date when the document becomes valid.
ValidUntil	Date until which the document remains valid.
Confidentiality	The level of secrecy of the document.
Status	The current status of the document. Examples of status values that might be used for a document information status include: DRAFT, FINAL DRAFT, FINAL REVISION.

TABLE 4.11

Extra Attributes About Legal Documents to Be Included in Description Attribute of "IfcDocumentInformation" Entity

Legal Document	Extra Information
Planning Permit	Permit Number, possible conditions and requirements.
Certification	Plan Number of surveyor's subdivision plan version.
Statement of Compliance	Council Reference Number, Office of Titles Plan Number, Plan Number of surveyor's subdivision plan version.
Engineering Plan	Engineering Plan Number.
Title	Volume Number, Folio Number, Security Number.
Parent Title	Volume Number, Folio Number.
Mortgage	Reference Number, Mortgagee Name, Mortgage Date.
Caveat	Reference Number, Caveator, Caveat Date.
Encumbrance	Reference Number and Encumbrance Date.

(see Figure 4.11a). For instance, "IfcRelAssociatesDocument" can be used to relate the title document to its owner and lot. The RelatingDocument attribute should refer to the title document information. Here, the RelatedObjects attribute should include both the IfcActor and IfcSpatialZone entities which are respectively used to model an owner and a multi-part volumetric lot.

Some legal documents are exchanged between land surveyors and city councils, as was explained in Section 2.3. These include the planning permit,

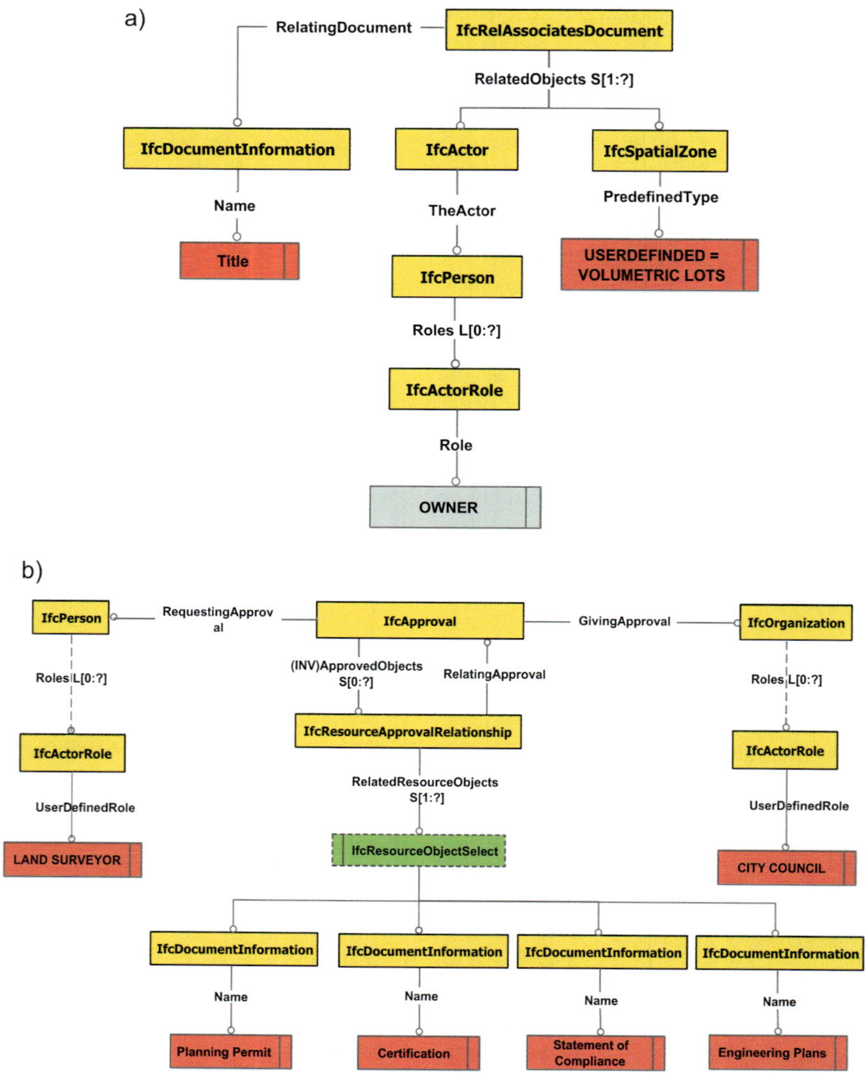

FIGURE 4.11
(a) Associating the title document to its owner and volumetric lot legal interest; (b) Modeling legal documents exchanged between city council and land surveyor within the IFC schema.

TABLE 4.12

Attributes Used for Approval of Legal Documents (Extracted from ISO16739 2013)

Attribute	Description
Identifier	A unique identifier for the approval of the legal document.
Name	Name given to the approval. For example, approval of the statement of compliance.
Description	A general textual description of the legal document that is being approved.
TimeOfApproval	Date and time of the decision on approval of the legal document.
Status	The result or current status of the approval of the legal document. Possible values could be Requested, Processed, Approved, Not Approved.
Qualifier	Textual description of special constraints or conditions for the approval. For example, conditions and requirements of a planning permit can also be included here.
RequestingApproval	The actor that is acting in the role of requesting the approval. For this research, "LAND SURVEYOR" actor should be assigned to this attribute.
GivingApproval	The actor that is acting in the role specified for giving the approval. "CITY COUNCIL" actor should be allocated to this attribute here.
ApprovedResources	The set of relationships associated with resource objects which are under approval process. In the context of this research, it should be associated with legal documents (instances of "IfcDocumentInformation" entity) via "IfcResourceApprovalRelationship" entity.

certification, statement of compliance, and engineering plan. The IfcApproval entity can be used to define data elements used in the approval process of these documents (see Figure 4.11b). Those attributes of the IfcApproval entity, which are essential in the context of urban land administration, are described in Table 4.12.

4.8 Administrative Information

As mentioned in Section 2.4.5, there are two types of administrative data elements: the survey header and plan information. The survey header provides information relating to the location of the development such as the geographic coordinates and postal address, which has already been provided as attributes in the IfcSite entity (see Table 4.13). Plan information can be added as a property set which is assigned to the IfcProject entity (see Figure 4.12). This property set is illustrated in Table 4.14.

TABLE 4.13

Attributes of "IfcSite" for Managing Survey Header Information (Extracted from ISO16739 2013)

Attribute	Data Type
RefLatitude	IfcCompoundPlaneAngleMeasure
RefLongitude	IfcCompoundPlaneAngleMeasure
RefElevation	IfcLengthMeasure
SiteAddress	IfcPostalAddress

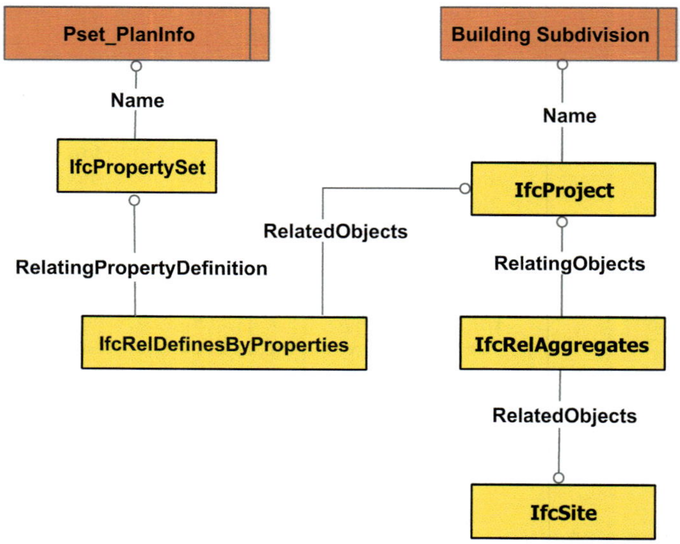

FIGURE 4.12
Modeling administrative information using the IfcSite and IfcProject entities.

TABLE 4.14

Proposed Property Set for Incorporating Plan Information Into "IfcProject"

Property Set Name	Pset_PlanInfo	
Name	**Property Type**	**Data Type**
PlanNo	IfcPropertySingleValue	IfcLable
StageNo	IfcPropertySingleValue	IfcInteger
LotsNo	IfcPropertySingleValue	IfcInteger
HeadOfPower	IfcPropertySingleValue	IfcLable
SurveyFormat	IfcPropertyEnumeratedValue	IfcLabel
Status	IfcPropertySingleValue	IfcText
AdministrativeDate	IfcPropertySingleValue	IfcDateTime

4.9 General Guidelines for Realizing BIM-Driven Digital Urban Land Administration

Although urban land administration practices are always dependent on the jurisdictional settings of a country or state, there are three steps to transform these practices into a BIM-driven 3D digital data environment:

- Identify data requirements for urban land administration: The first step is to conduct a requirements analysis to help identify the data elements required for urban land administration. The knowledge acquired from the investigation of current urban land administration practices can be used to define data requirements. In Chapter 2, these data elements are identified for the Victorian State of Australia. Other countries or states can adopt a similar approach of investigating current practices for urban land administration, which would subsequently help them explicate the major elements used for managing legal information and processes in complex urban environments.
- Design and develop an extension to the IFC standard: In this step, an extended data model of the IFC standard should be designed to meet the data requirements identified in the previous step. First, suitable and relevant IFC entities to be used in the proposed extension should be identified (see Chapter 3). Subsequently, in line with the logic of the IFC standard, the relevant IFC concepts will be extended to model land administration information with as minimal change as possible in the structure of IFC. The adopted approach to extending the IFC standard includes two main parts: property sets and user-defined values. In this chapter, it was explained how these extension mechanisms can be used to incorporate land administration data elements into the IFC standard.
- Implementation of prototype models: The final step is to implement the prototype BIM models to prove the viability of the developed extension of the IFC standard for 3D digital urban land administration. Suitable multi-story developments, which include various types of legal interests and boundaries, should be selected to provide a suitable demonstration of the extended IFC concepts. Chapter 5 provides some examples of the implemented prototype models to realize BIM-driven 3D digital urban land administration.

4.10 Concluding Remarks

In this chapter, an extension of the IFC standard was proposed to support the digital management of 3D data resulting from urban land administration

TABLE 4.15

A Summary of Addressing Legal Data Elements in the Proposed IFC Extension

Legal Data Element	Modeling the Legal Data Element in IFC Extension
Volumetric Lot	IFC entities: IfcSpatialZone, IfcSpace, and IfcZone Property Sets: Pset_LegalInterestCommon, and Pset_Lot
Land Lot	IFC entities: IfcSpatialZone, and IfcSite Property Sets: Pset_LegalInterestCommon, Pset_Lot
Common Property	IFC entities: IfcSpatialZone, IfcZone, IfcSpace, IfcExternalSpatialElement, IfcSite, and concrete subtypes of IfcElement Property Sets: Pset_LegalInterestCommon, and Pset_CommonProperty
Road	IFC entities: IfcSpatialZone, and IfcSite Property Sets: Pset_LegalInterestCommon
Reserve	IFC entities: IfcSpatialZone, and IfcSite Property Sets: Pset_LegalInterestCommon
Easement	IFC entities: IfcSpatialZone, IfcExternalSpatialElement, IfcSite and IfcDistributionElement Property Sets: Pset_LegalInterestCommon, Pset_EasementOr Restriction, Pset_Easement
Restriction	IFC entities: IfcSpatialZone, IfcExternalSpatialElement, IfcSite, and IfcBuildingElement Property Sets: Pset_LegalInterestCommon, Pset_ EasementOrRestriction, and Pset_Restriction
Depth Limitation	IFC entities: IfcSpatialZone, IfcExternalSpatialElement Property Sets: Pset_LegalInterestCommon, andPset_DepthLimitationOrAirspace
Airspace	IFC entities: IfcSpatialZone, IfcExternalSpatialElement Property Sets: Pset_LegalInterestCommon, and Pset_DepthLimitationOrAirspace
Internal Boundary	IfcRelSpacBoundary.RelatedBuildingElement = IfcBuildingElement IfcRelSpacBoundary.PhysicalOrVirtualBoundary = PHYSICAL IfcRelSpacBoundary.InternalOrExternalBoundary = INTERNAL
Median Boundary	IfcRelSpacBoundary.RelatedBuildingElement = IfcBuildingElement IfcRelSpacBoundary.PhysicalOrVirtualBoundary = PHYSICAL IfcRelSpacBoundary.InternalOrExternalBoundary = MEDIAN
External Boundary	IfcRelSpacBoundary.RelatedBuildingElement = IfcBuildingElement IfcRelSpacBoundary.PhysicalOrVirtualBoundary = PHYSICAL IfcRelSpacBoundary.InternalOrExternalBoundary = EXTERNAL
Other Boundary	IfcRelSpacBoundary.RelatedBuildingElement = IfcBuildingElement IfcRelSpacBoundary.PhysicalOrVirtualBoundary = PHYSICAL IfcRelSpacBoundary.InternalOrExternalBoundary = USERDEFINED
Projected Boundary	It is typically dependent on its corresponding building boundary. If it is modeled independently, then the following are applied: IfcRelSpacBoundary.RelatedBuildingElement = IfcVirtualElement IfcRelSpacBoundary.PhysicalOrVirtualBoundary = VIRTUAL

(*Continued*)

TABLE 4.15 (CONTINUED)

A Summary of Addressing Legal Data Elements in the Proposed IFC Extension

Legal Data Element	Modeling the Legal Data Element in IFC Extension
Ambulatory Boundary	IfcRelSpacBoundary.RelatedBuildingElement = IfcGeographicElement IfcRelSpacBoundary.PhysicalOrVirtualBoundary = PHYSICAL IfcRelSpacBoundary.InternalOrExternalBoundary = WATER_EXTERNAL
Fixed Boundary	IfcRelSpacBoundary.RelatedBuildingElement = IfcVirtualElement IfcRelSpacBoundary.PhysicalOrVirtualBoundary = VIRTUAL
Geometry of Boundaries	For boundaries in 2D space: IfcConnectionCurveGeometry should refer to IfcEdgeCurve which includes both topology and geometry of the boundary line For boundaries in 3D space: IfcConnectionSurfaceGeometry should refer to IfcFaceSurface which includes both topology and geometry of the boundary surface
Land surveyor	IFC entity: IfcPerson IfcActorRole.UserDefinedRole = DELINEATE
Owner	IFC entity: IfcPerson, IfcOrganization IfcActorRole.UserDefinedRole = OWN
Owners Corporation Manager	IFC entity: IfcPersonAndOrganization IfcActorRole.UserDefinedRole = MANAGE
Land registry	IFC entity: IfcOrganization IfcActorRole.UserDefinedRole = REGISTER
City Council	IFC entity: IfcOrganization IfcActorRole.UserDefinedRole = CERTIFY, OWN
Referral Authority	IFC entity: IfcOrganization IfcActorRole.UserDefinedRole = CONTROL, OWN
Developer	IFC entities: IfcPerson, IfcOrganization IfcActorRole.UserDefinedRole = DEVELOP
Legal Documents	IFC entities: IfcDocumentInformation, IfcApproval "Description" attribute of IfcDocumentInformation
Survey Header	IFC entity: IfcSite
Plan Information	IFC entity: IfcProject Property set: Pset_PlanInfo

practices. The culmination of this chapter was focused on extending the concept of the spatial and physical IFC elements required to lay the foundation for managing legal arrangements in urban land administration. Additionally, relevant geometric, topologic, and semantic IFC entities for modeling legal boundaries in the BIM environment were explained. The chapter also described how information relating to land administration actors and legal documents can be defined inside the IFC standard. Table 4.15 provides a summary of modeling legal data elements using the relevant IFC concepts.

Adopting an IFC-based approach underpins open and interoperable BIM environments for supporting 3D digital urban land administration. This approach would be considered as a fundamental step towards developing and implementing a BIM-driven urban land administration, which is specifically tailored to meet the 3D data requirements elicited from current practices in a particular jurisdiction. To showcase the viability of the proposed extension of the IFC standard for 3D digital urban land administration, prototype BIM models for simple and complex urban developments will be presented in the next chapter.

Reference

ISO16739. 2013. "Industry Foundation Classes (IFC) for Data Sharing in the Construction and Facility Management Industries." buildingSMART. Geneva, Switzerland: International Organization for Standardization.

5

Implementing BIM Models—From Simple to Complex Urban Developments

5.1 Introduction

An extension of the IFC standard was designed and presented in Chapter 4 to lay the foundation for digitalizing urban land administration practices within the BIM environment. To demonstrate the viability of the proposed IFC extension, this chapter presents simple and complex case studies used for implementing BIM models. These developments include common types of 3D legal interests such as volumetric lots, common properties, and easements. In addition, various types of legal boundaries have been defined in different parts of the developments. Therefore, the case studies are good choices for the representation and communication of various legal interests and boundary types in the digital environment of BIM.

As will be described in the following sections, the procedure for implementing prototype BIM models comprises three main steps:

- Constructing BIM models: For new buildings, BIM models are typically constructed from the early stages of the building development (see Section 5.2). These architectural BIM models can be also transformed into as-built models by conducting 3D surveying methods (see Section 5.3). For old buildings, 3D surveying data can be also used to construct as-built BIM models.
- Enrichment of BIM models: Either the design or as-built BIM model of a building can be enriched with the additional data used for the purpose of urban land administration (see Section 5.4). The IFC extension described in Chapter 4 plays an underpinning role in this enrichment.
- Visualization of enriched BIM models: The enriched BIM models are visually represented to showcase their capabilities in enabling 3D digital urban land administration (see Section 5.5).

5.2 Constructing BIM Models

BIM authoring tools are normally used to construct the BIM models. Some examples of these tools are Archicad, Autodesk Revit, Vectorworks, Bentley AECOsim, and so on. The presented models in this chapter are authored using Autodesk Revit software, as it represents one of the more well-known BIM authoring tools utilized in the AEC industry for constructing 3D BIM models of complex developments (Kurul et al. 2013). However, the steps explained here can be applied to other BIM-authoring tools by making minor changes.

First, 2D architectural floorplans for each level and cross-section diagrams in a CAD format are imported into the BIM-authoring tool. These CAD plans include a 2D representation for the architectural elements of the development. Subsequently, various 3D building elements, such as doors, walls, and windows, are created by extruding their footprints. The subdivision plan of the development was then used to delineate different types of legal boundaries. The correctness of these boundaries is checked and verified by consulting the land surveyor who prepared the subdivision plan of the development.

For legal boundaries referencing building structures, Revit, which is a well-known example of BIM authoring tools, has the capability to automatically create the boundary by checking the "Room Bounding" attribute. In addition, the location of each building boundary is set to an interior, median, exterior, or other position inside the physical structure as specified in the subdivision plan of the development. Fixed boundaries are modeled using the "Room Separator" tool in Revit. The delineated legal boundaries specify the 3D delimitation of various parts of lots, common properties, roads, and reserves. Once legal boundaries are delineated, each legal space is automatically constructed by applying either the "Room" or "Space" tool in Revit. After defining all legal spaces, the prototype BIM model was converted from the proprietary format of the BIM-authoring tool into the open IFC format. The summary of steps for constructing the prototype BIM models is outlined in Figure 5.1.

5.3 Constructing BIM Models from 3D Surveying Data

While BIM processes can easily be considered for new buildings, the majority of existing buildings are, however, not yet maintained within the BIM environment. As part of the rapid developments in BIM research, 3D surveying methods have been explored and reviewed for sourcing the data required in a BIM model. If building information is insufficient for the required functionalities, techniques of 3D data capture or survey are applied.

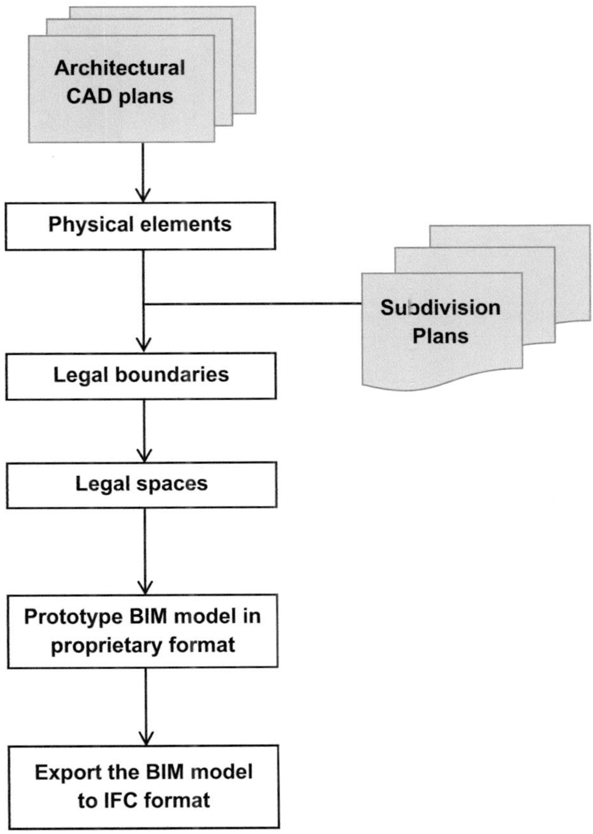

FIGURE 5.1
Steps for constructing BIM models in a BIM-authoring environment.

Traditionally, 3D building models have been generated automatically from 3D point clouds using laser-scanning, or from image-based methods using aerial imagery with a focus on the facade and roof data of the buildings. While significant developments in 3D data acquisition techniques, visualization, and computer power have facilitated the creation of 3D models with increasing levels of detail, accuracy, and scale, the resulting models are generally primarily focused on the facade and roof information only.

Adopting a methodology that integrates both the internal and external attributes of buildings has the potential to address the shortcomings of current processes that focus on reconstructing the outside of a building or indoor mapping in silos. The 3D information that is required relates to (i) the external attributes of the building, including the facade, roof, and texture; (ii) the internal attributes, including 2D indoor floorplans, building exits, utilities, etc. (see Figure 5.2).

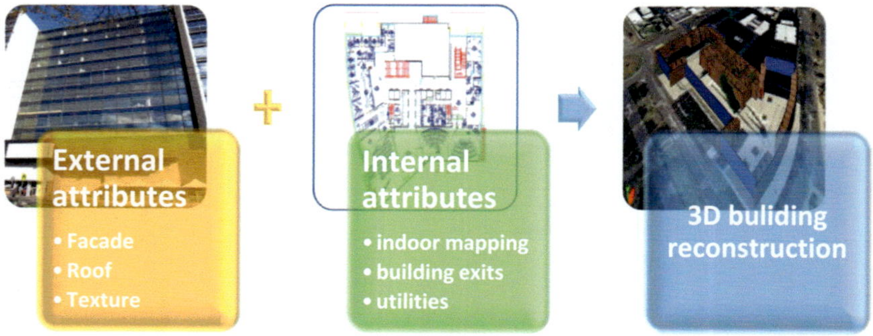

FIGURE 5.2
Combined external and internal building attributes for complete 3D building reconstruction.

The fusion of 3D data will result in the culmination of rich and detailed information describing both the external and internal attributes of building, enabling a more complete reconstruction of 3D building models required for a BIM.

There are various techniques for acquisition of exterior attributes of buildings which are used for 3D city model generation. During the last decade, laser-scanning, both airborne and terrestrial, has been one of the most significant advancements in acquiring 3D data. Airborne Light Detection and Ranging (LiDAR) is now a standard technology for 3D data acquisition, integrating with GNSS for direct geo-referencing capabilities. There are two basic types of LiDAR sensors, namely ground-based sensors (both fixed and moving, i.e. mobile mapping from car vehicles) and airborne sensors, such as those attached to satellites and aircraft.

Close-range photogrammetry is a measurement technology which is used to acquire 3D spatial information about an object, captured using digital cameras. By taking multiple convergent photographs of an object using an off-the-shelf camera and processing the imagery using photogrammetric techniques, it is possible to create a point cloud with high accuracy 3D point information. The imagery taken can also be used to extract texture information to create photorealistic 3D models. There are a number of advantages to using close-range photogrammetry for building reconstruction including low cost, fast data capture, high accuracy, minimal data post-processing, and photorealism of results (Remondino and El-Hakim 2006).

Additionally, there are other approaches that recently have been utilized for 3D model reconstruction, such as Unmanned Aerial Vehicles (UAV) which will be discussed in the next sections.

3D reconstruction of buildings relating to the physical objects is fundamentally 3D modeling for geometric object reconstruction, where each feature of the object (i.e. roof, facade, wall, etc.) can be identified and classified. Most of the literature in this area reports on the external attributes of the building, i.e. reconstructing the facade and roof of the building only.

There are a number of available methodologies that have been developed to accomplish this, including photogrammetry, computer vision, computer graphics, and laser-scanning. Many applications involve the creation of visually-pleasing models with great photorealism, but little geometric quality, such as in the animation or the gaming industry. This book focuses on the methodologies which yield geometric results with rich geospatial attributes that can be implemented in the creation of BIM models used in effective decision-making.

3D reconstruction methods can be classified into two main categories:

- Image-based modeling via passive sensors (using imagery) such as close-range photogrammetry, high-resolution satellite imagery, architectural plans, Unmanned Aerial Systems (UAS) imagery.
- Range-based modeling via active sensors, which includes terrestrial laser-scanning, LiDAR, and mobile mapping.

Image-based modeling is defined as a complete process that starts with image acquisition and ends with an interactive 3D virtual model camera photographs using close-range photogrammetry techniques (Remondino and El-Hakim 2006). Image-based solutions have been widely researched and can include input images from single images or from multiple image networks using aerial imagery, satellite imagery, or off-the-shelf cameras. Advancements in digital photogrammetry has meant that image-matching algorithms can be used to produce 3D models fit to compete with laser-scanning with respect to metric accuracy. The advantages of image-based solutions include low cost, portability, and short collection time, as well as delivering 3D models with a high level of accuracy and photorealism (Jazayeri, Rajabifard and Kalantari 2014).

Range-based modeling such as LiDAR, directly captures 3D information, providing a rich and dense dataset. This can be used for the automatic reconstruction of urban building models, or more commonly for the extraction of the rooftop or facade of 3D buildings (Keqi, Jianhua and Chen 2009; Oude Elberink and Vosselman 2011). While this method is costly and data-intensive, it provides automatic data-processing options for the extraction of building dimensions and spatial locations with potential for spatial analysis. LiDAR has the advantages of high accuracy, rapid acquisition, and high resolution, which are required for large-scale elevation collection and object extraction. External building reconstruction can be achieved using LiDAR alone (Wang and Schenk 2000), or through the integration with aerial imagery (Chen et al. 2005; Chen and Teo 2004; Rottensteiner 2003) satellite imagery (Sohn and Dowman 2007), or close-range photogrammetry (Habib, Ghanma and Tait 2004).

Internal building reconstruction and indoor mapping has been gaining greater attention over the past few years. This trend is evident in a number of modeling datasets available, including indoor Google Maps which provides maps for indoor spaces such as airports, shopping centers, large retail stores, and transit stations. Such solutions, however, are generally 2D, with little integration capabilities with full 3D building models and indoor views. Moving from an external reconstruction to a building model with internal attributes provides a natural progression towards a complete solution.

Creating 3D indoor plans from 2D CAD files generally involves extruding the 2D information to visualize in 3D. The complication of this process is often underestimated, and the conversion from 2D to 3D is still a challenging problem today. If digital versions of plans exist, they are often only in PDF format where the paper plan has been scanned, with little or no functionality or opportunity to convert to 3D. Additionally, it is important to note that many building plans are often still paper-based. There is therefore no unique solution in place to process these plans, and a number of different algorithms are required for the varying processing steps, based on the format of the input data. These file types preserve the drawing information as 2D geometric primitives, making the conversion process easier, with the possibility of automated data-processing. If a user simply scans 2D plans into a digital format, the result is a raster image with no distinction between graphical symbols, utilities, wall lines, windows, and other building components. In order to make sense of this kind of information, a conversion system must rely on image-processing and pattern recognition techniques. CAD-based solutions therefore have an advantage, in that there is embedded semantic information and fewer ambiguities compared to image-processing techniques. However, this approach still currently requires considerable manual effort. While there are a number of automated solutions in place, fully automated error-free 3D model extrusion still remains a challenging area of research and practice.

5.3.1 BIM Models from Laser-Scanning Data

Using laser scanners to create point clouds when generating as-built information is standard practice in a range of applications, including building and facilities management. The acquisition of 3D point information using laser-scanning has led to significant research on the development of algorithms for processing 3D point cloud data. In the commercial realm, however, the level of automation of current software solutions for processing TLS data remains somewhat restricted.

In reality, there are a number of disadvantages to using laser-scanning methods, including high equipment cost, as well as difficulties in scanning reflective, dark, and transparent surfaces. In addition, there is a significant amount of post-processing required, including extensive data-processing and modeling steps, and current approaches have very minor levels of detail (LODs).

The automatic procedure of generating BIM objects from point cloud data generally consists of the following steps (Nguyen and Le 2013). In the first step, the data is pre-processed in order to make the modeling process more efficient. The point cloud is then represented as a set of raster images for 2D methods, consisting of a slice of the data, or any other prevalent information. When pre-processing for 3D methods, the point cloud is restructured in order to allow efficient neighborhood searches by implementing a voxel octree (Vo et al. 2015; Vosselman, Coenen, and Rottensteiner 2017). In the next steps, a classification of the segments is implemented by reasoning frameworks analyzing local and contextual information. Using machine learning techniques, the class labels such as walls and floors are computed (Wolf, Prankl, and Vincze 2015).

In order to align this reconstruction to the BIM realm, Bassier et al. (2018) add a further step which allows for the reconstruction of BIM geometry, specifically wall geometry. This final step is to convert the geometry to actual BIM objects. In their research, the IFC4 protocol is implemented, as it is compatible with most software. Instances of the IfcWallStandardCase class are generated to have access to the extensive parameter set provided for this class. A limitation, however, is that these objects have geometric constraints, such as verticality and uniform thickness. To address this limitation, an abstraction is made of the geometry of each wall to create the objects along the Z-direction with a uniform height and thickness. This results in a set of best fit generic parametric BIM objects that reflect the current state of the building.

To test this, Bassier et al. (2018) implemented their methodology on a multistory building. A total of 55 scans were acquired, which resulted in over half a billion 3D points. The accuracy of the reconstructed BIM objects was obtained by testing against the initial point cloud which serves as ground truth. Results of their experiments indicated that the used method is a fast and reliable reconstruction framework for unstructured point cloud data. Bassier et al. (2018) showed over 90 percent recall together with a high precision for the clustering of the wall segments, the reconstruction of the wall topology, and the detection of the partial rooms. In terms of accuracy, it is revealed that the majority of the wall boundaries are accurately reconstructed when comparing to the initial point cloud. The problem areas are major deviations near wall detailing, due to the abstractions made to create IfcWallStandardCase objects.

5.3.2 BIM Models from UAV Data

Over the past few years, lightweight UAVs have received increasing attention and been a subject of significant research in building reconstruction, due to their multi-sensory data collection capabilities and their ability to provide an agile and highly maneuverable robotic platform for the reality capture technologies. As a result, there has been an exponential growth in the volume of images and videos that are being recorded on construction sites and building sites around the world. An exciting outcome of this is that at a fraction of the cost of laser-scanning methods, it is now possible to acquire a very large volume of digital images that provides unprecedented opportunity to visually capture the actual status of construction sites.

Currently, most work documenting the as-built status of buildings stay at collecting visual data and updating BIM. The hundreds of images and videos that are captured during construction are generally not utilized without being properly localized with plan document and time. This provides an opportunity to create a mechanism that would localize a large amount of visual data with respect to BIM, in order to provide a useful as-built status with accurate location. There is a need for some sort of sensing mechanism that would enable an automatic information flow at a high frequency for capturing changes in real time.

Ham et al. (2016) have identified that in order to address the issues discussed, what the AEC industry needs to solve using big visual data are: (1) data collection and analytics—visual data captured at a high rate of frequency and its analytics that provide insights on project controls; (2) reporting—effectively capturing and reporting issues found on construction sites; and (3) communication—efficient and effective flow of information within a construction site and site-to-office.

The two specific challenges associated with the localization of images with respect to BIM include: (1) directly aligning images to BIM; and (2) aligning via reconstructing a 3D point cloud from image collections.

When it comes to the placement of time-lapse cameras, it is recommended that the UAV instrument should be at a considerably higher elevation than the building and site being recorded. The camera should also capture a wide field of view in order to minimize occlusions by foreground activities or other obstructions. Following these considerations can result in a better camera viewpoint and larger image surfaces for the construction monitoring purposes and efficient data-collection.

5.3.2.1 Image to BIM Alignment

There are a number of major considerations that need to be addressed when discussing the use of imagery and BIM. To begin with, it is imperative to perform a camera calibration that will in turn allow for automated registration and alignment. This can be done via automated feature detection matching between convergent geometric images. There needs to be a high number of images, recorded at fixed focus, with diverse camera roll angles in a strongly convergent imaging configuration.

Images and data from UAVs are typically calibrated, and therefore their location and their orientations are generally known. For this reason, it is possible to accurately localize the images and place them in the correct geometric location in the BIM. At present, however, it is still a challenging process to locate the exact views in BIM for a 2D to 3D comparison, due to inaccurate GPS coordinates, particularly in the vertical axis.

The image to BIM alignment is achieved by creating a point cloud and selecting correspondences between the point cloud and BIM, followed by a similarity transformation. After selecting a number of corresponding points, the two systems will align, and meaningful data regarding the 3D points and their corresponding BIM information can be extracted. If pre-surveyed control points are used, the process of matching corresponding points is made simpler, faster, and more accurate. Any issues with radial distortion, or any other distortion, are addressed in the camera calibration process.

5.3.2.2 Point Cloud to BIM Alignment

In the case of using camera-equipped UAVs, the GPS coordinates synced with images can bring point clouds into alignment with BIMs. This process requires BIMs to be in the world coordinate system (typically Universal Transverse Mercator [UTM]). There is currently a lot of commercial software that can perform this alignment automatically, with user-friendly tools that result in centimeter accuracy.

5.3.3 UAV Localization

Localizing UAVs is a critical element in autonomous navigation, where the UAV can autonomously define a trajectory through the environment in order to capture all the building data required, while also dealing with any possible unexpected obstacles. Simultaneous localization and mapping (SLAM) is a major area of research for autonomous navigation. SLAM can often have challenges associated with drift, especially when areas of interest have large baselines. Using BIM, as a priori geometric information, SLAM can potentially be used to correct or calibrate the position of the UAV with respect to BIM (Ham et al. 2016).

Ham et al. (2016) have shown case studies creating BIM models with images taken by a UAV for a hospital, university, and sports facility. The real-world cases were used to show how BIM can be sourced from UAV technologies to streamline communication among all project participants.

The results showed the approach suggested was efficient, safe, and cost-effective compared to conventional manual practices. The main limitation discussed was the technology gap between the integration of the UAV imagery into an automatically and continuously updated BIM (Hamledari et al. 2017)

Hamledari et al. (2017) developed a method to automatically incorporate the progress data into an IFC-based 4D BIM, and modify its schedule and progress information accordingly. This was achieved by analyzing object (IfcObject)-task (IfcTask) relationships, modifying the task–task hierarchy, and updating schedule and relationship-related data instances in the IFC data model.

The culmination of this means that the user can implement the generated 4D BIM for later updates, visualization, communication of progress, and

4D planning. It is a semi-automatic process, meaning that there is some user involvement required for including data capture at the site and providing the input 4D BIM. The progress assessment and 4D BIM updating process are fully automated. Implementing the IFC data exchange format eliminates reliance on commercial software packages and also enables the interoperable exchange of information throughout the building lifecycle.

5.3.4 BIM Models from LiDAR, Photogrammetry, and 2D Floorplans

3D building reconstruction from LiDAR point clouds has been a challenging research topic for decades in photogrammetry and computer vision, and will continue to be in future as long as acquisition systems are improving and requirements for 3D city models are increasing. Recent developments in LiDAR sensor technology, with the increasing capabilities of sensor data, generates higher demand for highly detailed and realistic 3D building models. Many approaches have been reported in the literature, and overviews can be found in Brenner (2005); Schnabel et al. (2008); Vosselman and Dijkman (2001); and Shan and Toth (2010).

Original LiDAR data is generally very dense and detailed, but with highly accurate 3D point information. Close-range photogrammetry results produce less detailed point clouds but capture texture information. The imagery from the photographs are generally high-resolution pictures with a high-level of visual information. 3D modeling using a combination of image and range data is a sensor fusion methodology that takes the strengths from each approach, in order to overcome their limitations.

A test was run at the University of Melbourne to reconstruct a multi-story building using only LiDAR, close-range photogrammetry, and 2D floorplans. The results of the LiDAR model are shown in Figure 5.3a.

Using an off-the-shelf consumer grade camera, 18 photos of the building were taken, and using a multi-ray image matching process, a reconstruction of the building was generated in the form of a 3D point cloud. This involves

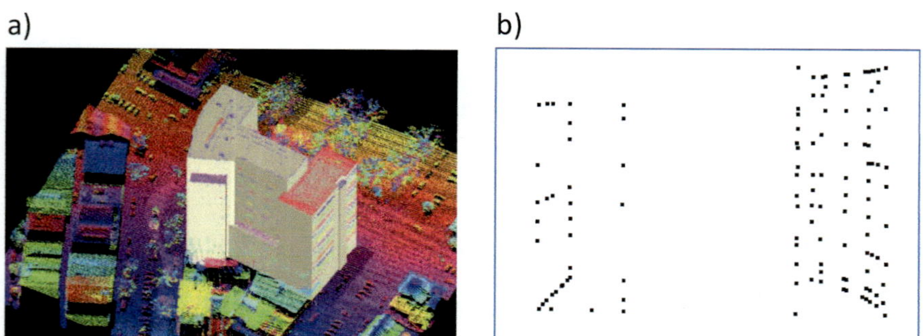

FIGURE 5.3
(a) Reconstructed 3D model from LiDAR data; (b) 3D point cloud of building.

first orienting the network, calibrating the camera, and then determining the coordinates of the points describing the building in 3D space through image point correspondence, a form of feature-based matching. The results of the matching process are illustrated in Figure 5.3b.

Once the 3D point cloud was determined, the texture patches between the 3D points could be generated. The building was textured via back-projection of the 3D points determined by the user into the most appropriate images of the network, these being selected using view-dependence criteria and radiometric considerations. The texture patches were ortho-rectified images mapped on to each surface of the building. The final texture patches, as well as the resulting point cloud comprising 3D points, formed the final output from the close-range photogrammetric processing.

The generation of 3D indoor floorplans required 2D CAD files of each floor of the building shown in Figure 5.4a. The 2D CAD files were imported into the program Autodesk Revit, software designed for architects, particularly for applications relating to BIM.

The 2D floorplans were transformed into 3D plans manually, by assigning each of the walls, doors, and windows their height, width, and depth from a list of available options which could be customized in Autodesk Revit. This was a straightforward procedure and allowed for an accurate reconstruction of the 3D information for all floors of the building.

The software Blender was used to integrate the 3D model generated by Autodesk Revit with LiDAR and close-range photogrammetry. Blender itself does not support importing and exporting IFC files, but a plug-in* was available to provide this functionality. However, this did not convert any material information such as surface colors. Therefore, colors, transparency, and texture were added to the 3D model (see Figure 5.4b). All textures were added to the 3D model and the finalized 3D model was exported to the Collada format.

In order to visualize this 3D model, a 3D web-based visualization system was developed using WebGL technology, as illustrated in Figure 5.4c. WebGL is a cross-platform, royalty-free web standard based on OpenGL, and provides users with 3D models using a canvas element, a container for graphics, in HTML 5. WebGL brings plug-in-free 3D to the web and major browsers.

Using the developed approach, it is possible to reconstruct both the internal and external attributes of a building in 3D through the fusion of LiDAR data, close-range photogrammetry, and 2D floorplans, and visualize the output on a web-based platform with added functionality. The results from the case study presented here indicate that the developed methodology generates photorealistic and geometrically accurate 3D buildings with complete detail in a geo-referenced environment.

* http://ifcopenshell.org/ifcblender.html

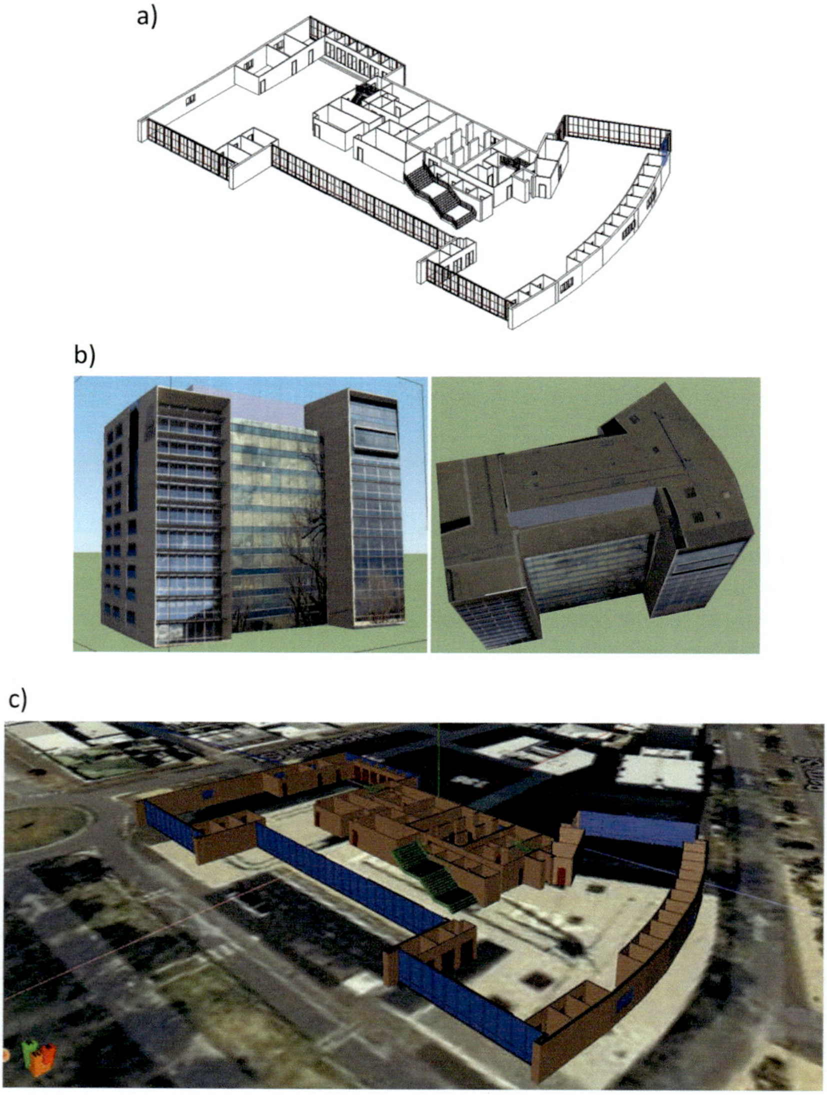

FIGURE 5.4
(a) A snapshot of a 3D plan from a 2D CAD floorplan; (b) 3D building model with realistic texture; (c) Geo-referenced 3D plans.

5.4 Enriching the BIM Models for Urban Land Administration

After creating either design or as-built BIM models, these models are enriched with legal data elements required in urban land administration practices. Therefore, the exported BIM model in the IFC format is enriched according to the proposed extension of the IFC standard, as described in Chapter 4. Application Programming Interfaces (API) for handing IFC files can be used to automate the process of enrichment. One open source API is xBIM (eXtensible Building Information Modeling) toolkit. This API provides various capabilities in terms of manipulating IFC files such as reading, enriching, writing and querying these files. The enrichment of the BIM model includes the following steps:

- Defining spatial arrangements of each legal interest
- Assignment of attributes as property sets to each legal interest
- Assignment of land administration actors
- Incorporating and referencing information about legal documents
- Adding administrative information to the IFC file

5.5 Visualization of BIM Models

After finalizing the enriched BIM model, it is imported into the Solibri Model Viewer (SMV) software, which provides 3D visualization of IFC files. SMV has the capability to show legal interests and boundaries in a more user-friendly way. Also, all attributes and property sets assigned to legal interests can be viewed in the SMV environment. More importantly, SMV clearly highlights and represents boundaries of legal spaces.

In this section, three prototype BIM models for urban developments with different levels of spatial complexities will be represented to showcase the feasibility of the proposed IFC extension for BIM-driven 3D digital urban land administration. These prototype models include a simple two-story building (development A), a partly complex four-story development (development B), and a complex development comprising a six-level apartment and three townhouses (development C). More specifically, the examples from the prototype models will show how each legal data element can be represented in the 3D digital data environment of BIM.

5.5.1 Primary Legal Interests

Primary legal interests include lot, common property, road, and reserve in urban land administration practices in the Victorian jurisdiction (see Section 2.4.1.1). This section provides examples of primary legal interests within prototype BIM models. A normal example of a single part lot can be a house unit or an apartment unit. Figure 5.5a shows an example of a house unit in development A, while Figure 5.5b provides an example of an apartment unit in development B. A typical example of a multi-part lot is

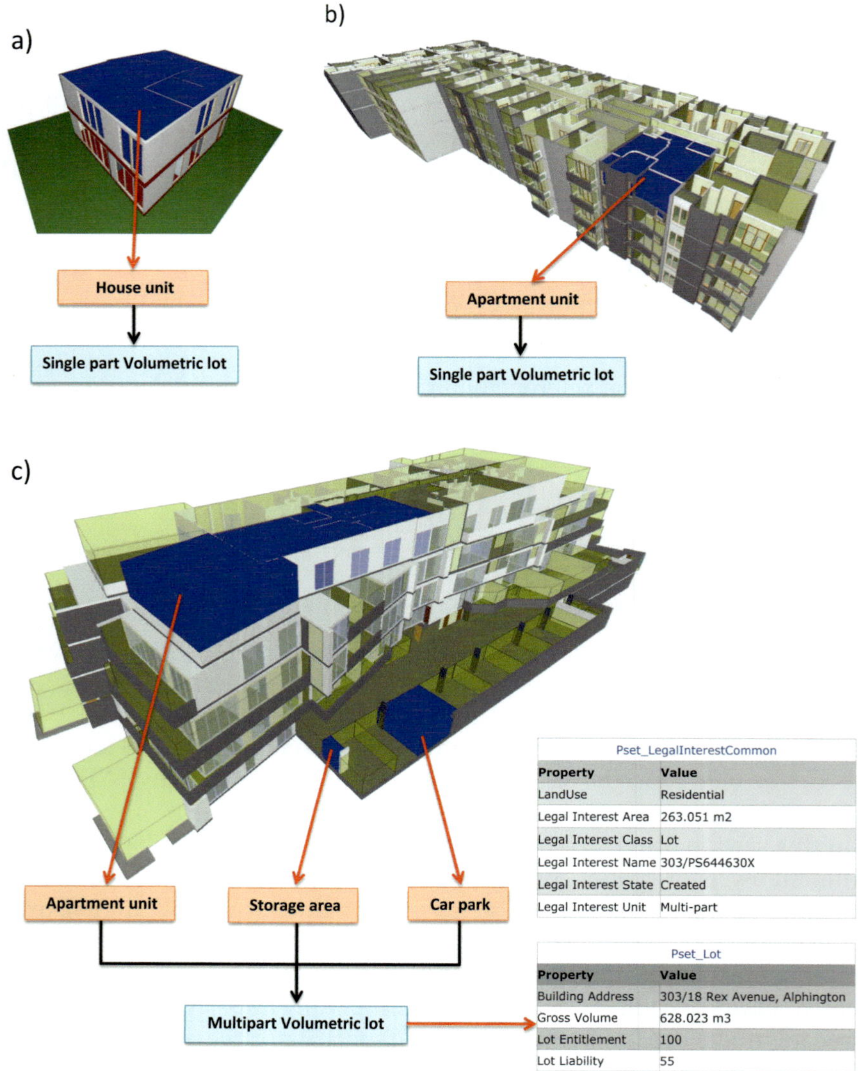

FIGURE 5.5
Examples of volumetric lot in BIM models; (a) Development A; (b) Development B; (c) Development C.

composed of one apartment unit, one car park, and one storage area (see Figure 5.5c). Additionally, for each volumetric lot, various attributes defined as property sets can be accessible through the attribute manager of SMV. For instance, as illustrated in Figure 5.5c, the liability and entitlement of LOT-303 are 220 and 100 respectively, which can be found only by clicking on the specific zone defined for this lot. This makes it easier for users to find information about each volumetric lot defined inside complex developments.

Common properties are typically composed of multiple parts, even in simple developments. In development A, the common property is composed of the external spaces around the house and the ceiling between two floors (see Figure 5.6a). Another example of the common property is composed of only internal legal spaces, as illustrated in Figure 5.6b. A more complex example of the common property includes not only various communal spaces, such as corridors, stairs, and elevators but also a wide range of physical structures such as walls, columns, and slabs (see Figure 5.6c). Parts of communal

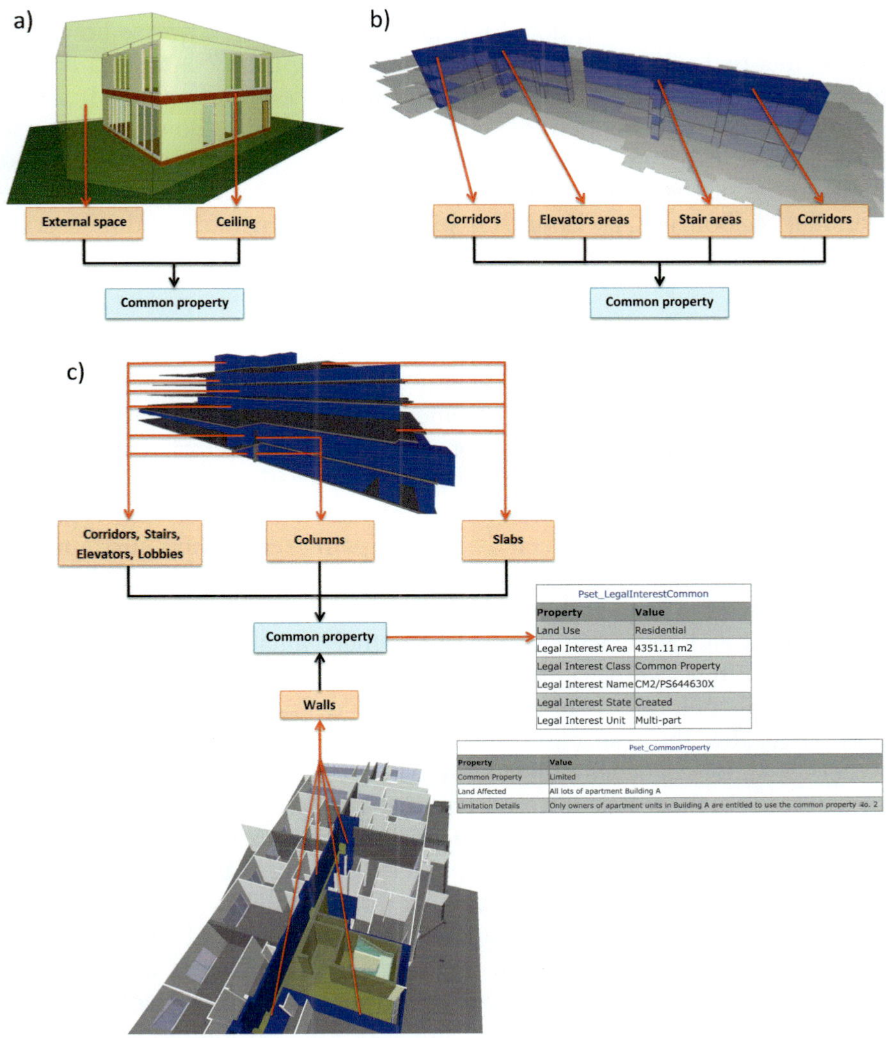

FIGURE 5.6
Examples of common property in BIM models; (a) Development A; (b) Development B; (c) Development C.

spaces passing through several levels of buildings can be assembled inside the common property spatial zone, making it easier to manage such complex legal arrangements. In addition, those building elements, such as walls and columns, defined as part of the common property spatial zone can be easily located inside the BIM environment. This implies that common property areas are more clearly communicated in the BIM environment since there is no abstraction of physical structures.

Among the case studies, only development C includes road and reserve legal interests. The spatial extent of these primary legal interests is defined as land parcels within the developments. These legal interests were created within the prototype BIM model for development C to showcase how they can be mapped within the BIM environment. Figure 5.7a shows a spatial zone of a road defined by land parcels. As mentioned in Chapter 4, IfcSite is the suitable entity for modeling land parcels. Figure 5.7b shows an example of the reserve legal interest. In this case, two parcels, each of which is defined by instantiating the IfcSite entity, constitute the reserve spatial zone. It can be seen that Melbourne City Council holds the legal ownership of this reserve. In addition, the "Gross Area" quantity was used to calculate the whole area of the reserve.

5.5.2 Secondary Legal Interests

Secondary legal interests include easement, restriction, depth limitation, and airspace in urban land administration practices in the Victorian jurisdiction (see Section 2.4.1.2). This section provides examples of secondary legal interests within prototype BIM models. An instance of the easement typically comprises utility elements, such as water pipelines, and their surrounding legal spaces. In Figure 5.8, the attribute manager of the SMV indicates that all lots on the plan are affected by the easement shown. In addition, the BIM environment can also facilitate access to other valuable attributes such as purpose, width, and the origin of the easement.

An example of a restriction can be spatially defined by those parts of the land parcel that the restriction is applied to. As illustrated in Figure 5.9a, owners of all lots in development C shall not construct any building requiring sewerage in the restriction area. Another example for defining the spatial arrangement of a restriction can be based on some wall elements in development B to which the restriction is applied (see Figure 5.9b). For example, the restriction could specify a particular color for the walls.

In current practices, depth limitations are only described by textual information indicating the length of the depth limitation. However, the BIM environment provides the ability to visually represent the spatial extent of a depth limitation. Figure 5.10a shows an example of a depth limitation. Airspace legal interests can also be modeled by defining external spaces around the building. Figure 5.10b shows a specific instance of an airspace legal interest, which indicates the height limit of the building.

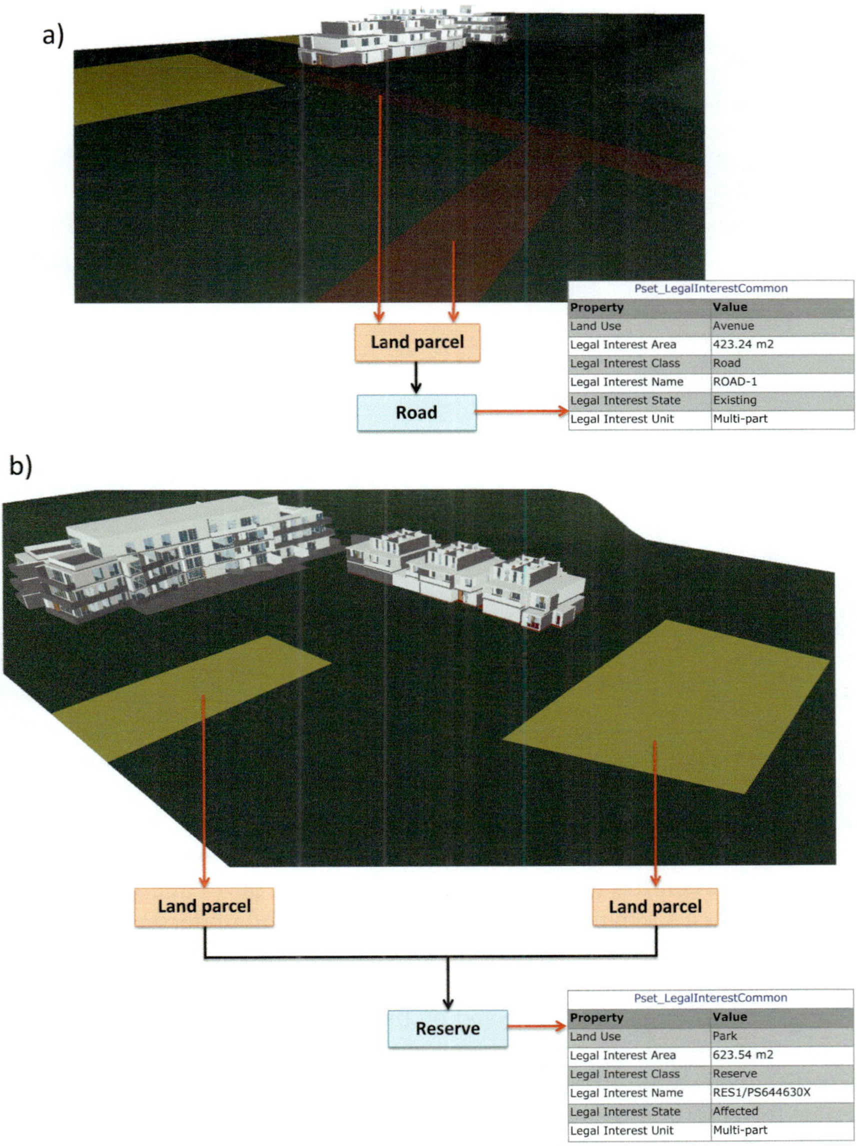

FIGURE 5.7
(a) An example of a road legal interest; (b) An example of a reserve legal interest.

5.5.3 Legal Boundaries

A wide range of legal boundaries was identified in Section 2.4.1.3. This section illustrates how the prototype BIM models can represent different types of legal boundaries. In Figure 5.11a, only the legal spaces for legal interests are shown, and it is hard to find the position of the boundary between two

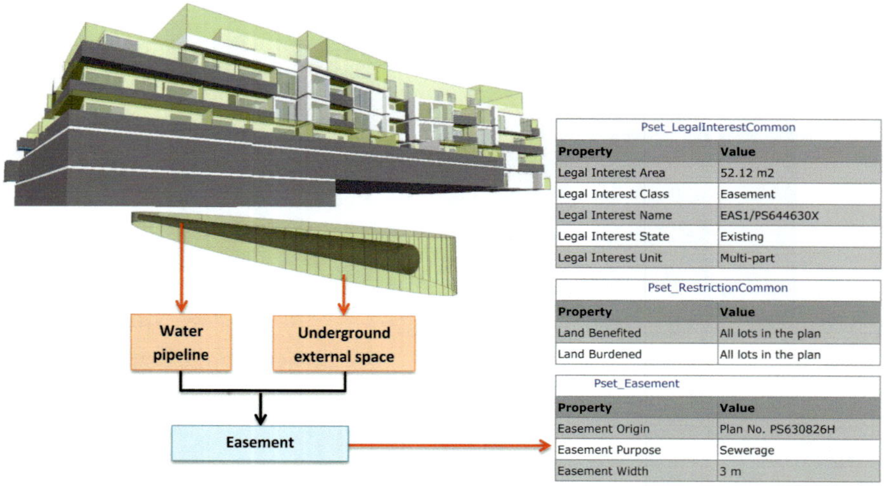

FIGURE 5.8
An example of the easement.

abutting legal spaces. This means that a purely legal representation of legal interests itself cannot adequately communicate the type and position of legal boundaries. However, incorporating physical objects facilitates communication of the boundaries. In the next paragraphs, it is discussed how a BIM environment would enhance the communication of various legal boundaries by improving on the visual representation.

Figure 5.11b represents two apartment units (No. 203 and 204) within the second level of an apartment building in development C, in which the interior boundary between the units was highlighted in the BIM environment. For a person without a specialist background knowledge and understanding of subdivision plans, it is difficult to interpret the location of the boundary. In contrast, it can be intuitively perceived inside the BIM environment that the boundaries of the apartment units are delineated by the internal surfaces of the wall. This would communicate that neither owner of the apartment units has any rights to the wall, which would subsequently be a part of the common property.

Some townhouses are represented in Figure 5.11c, which shows two abutting townhouses (No. 2 and 3) and the position of their common median boundary in a BIM environment. A land surveyor might define a median boundary through the use of an "M" annotation alongside the boundary line in the subdivision plan. However, the owners of these townhouses may have difficulty understanding this boundary. When represented in a BIM environment, this boundary is unambiguous, as shown in Figure 5.5c. Each owner can clearly understand that their legal interest is spatially extended up to half of the wall on their side.

As illustrated in Figure 5.11d, a typical example of the external boundary for townhouse No. 4 is highlighted in the BIM environment. The owner of townhouse No. 4 may not accurately interpret the location of this boundary

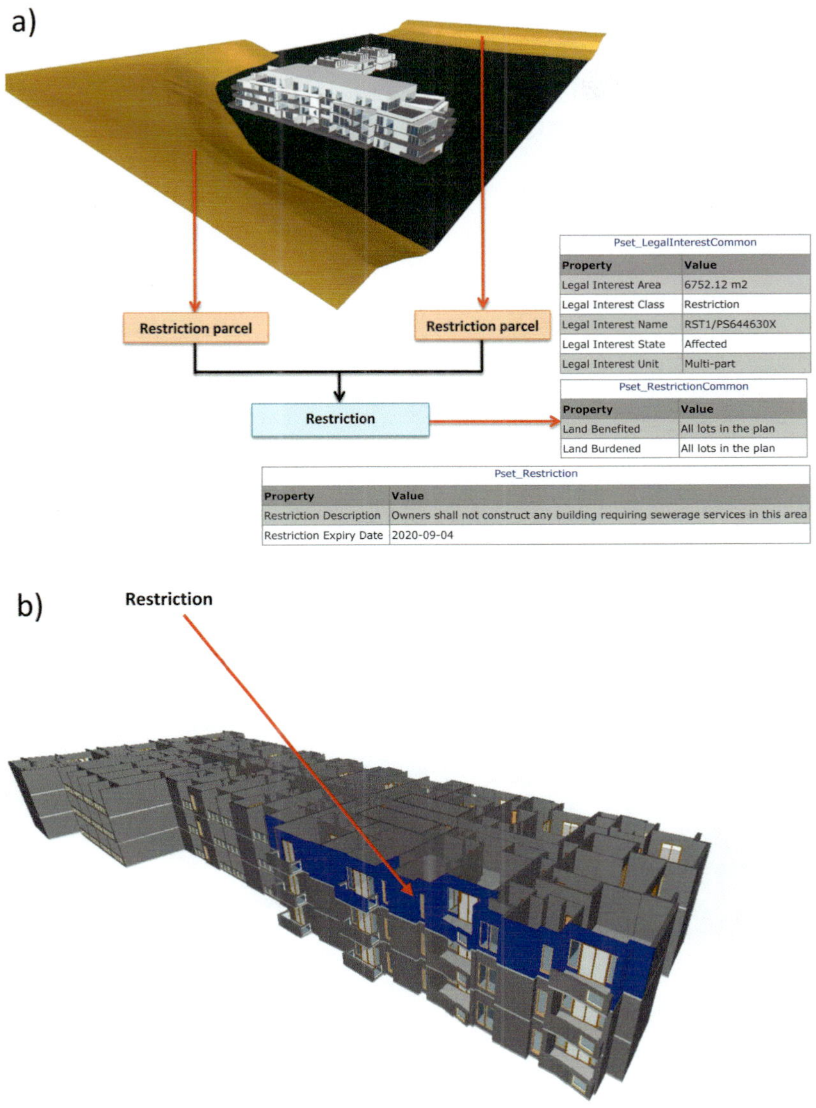

FIGURE 5.9
Examples of the restriction.

in a 2D plan. Mapping this boundary into the BIM environment, as illustrated in Figure 5.11d, could be more congruous with the visual perception of non-specialists. This would easily communicate that the building element associated with the boundary (in this case, the wall) completely belongs to townhouse No. 4.

For projected boundaries, the BIM environment can represent these automatically by extruding the physical boundary surfaces in both vertical and

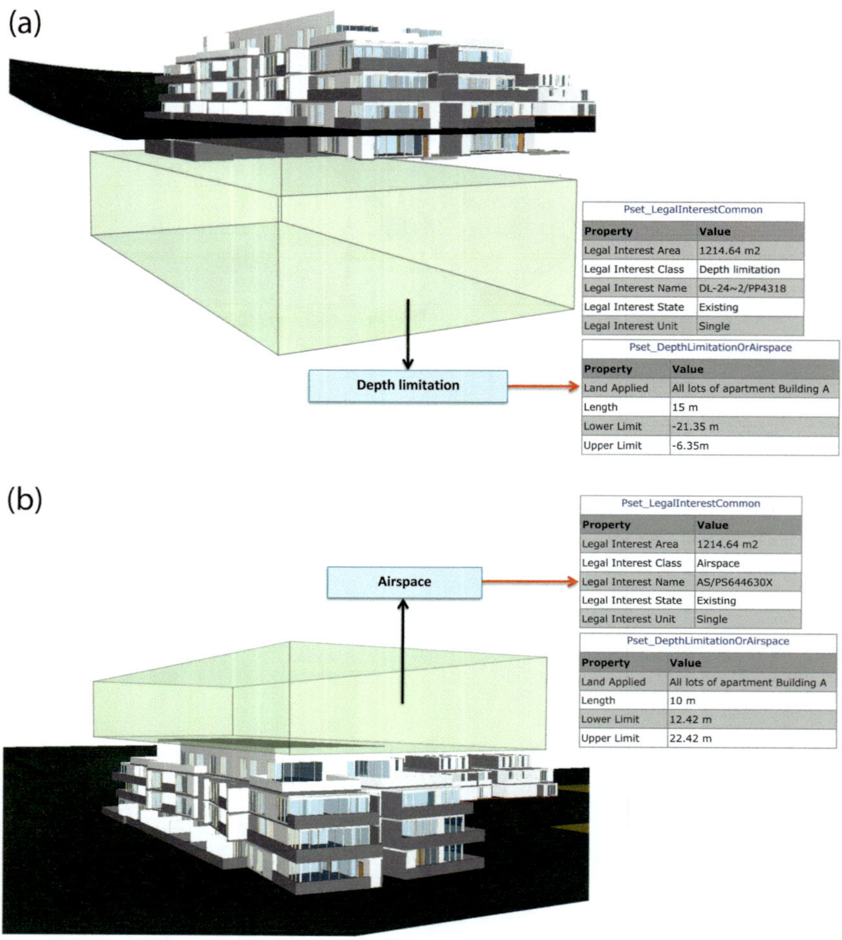

FIGURE 5.10
(a) An example of a depth limitation; (b) An example of airspace.

horizontal directions (see Figure 5.11e). In subdivision plans, these can be misinterpreted, similar to building boundaries, since their location is dependent on the position of their corresponding structural boundaries. However, modeling in a BIM environment can facilitate communication of these boundaries in the same way as structural boundaries.

Figure 5.11f shows how fixed boundaries associated with a car park area can be represented in the BIM environment. Interpretation of fixed boundaries in subdivision plans can also be difficult for owners. One typical issue with the fixed boundaries is that owners may only look at the floorplan of their car park area and think that the legal space of their car park is up to the ceiling. However, in cross-section diagrams, it is sometimes indicated that the vertical extension of the car park is only up to a height of 2 meters. The remaining space from a height of 2 meters to the ceiling is part of the common property

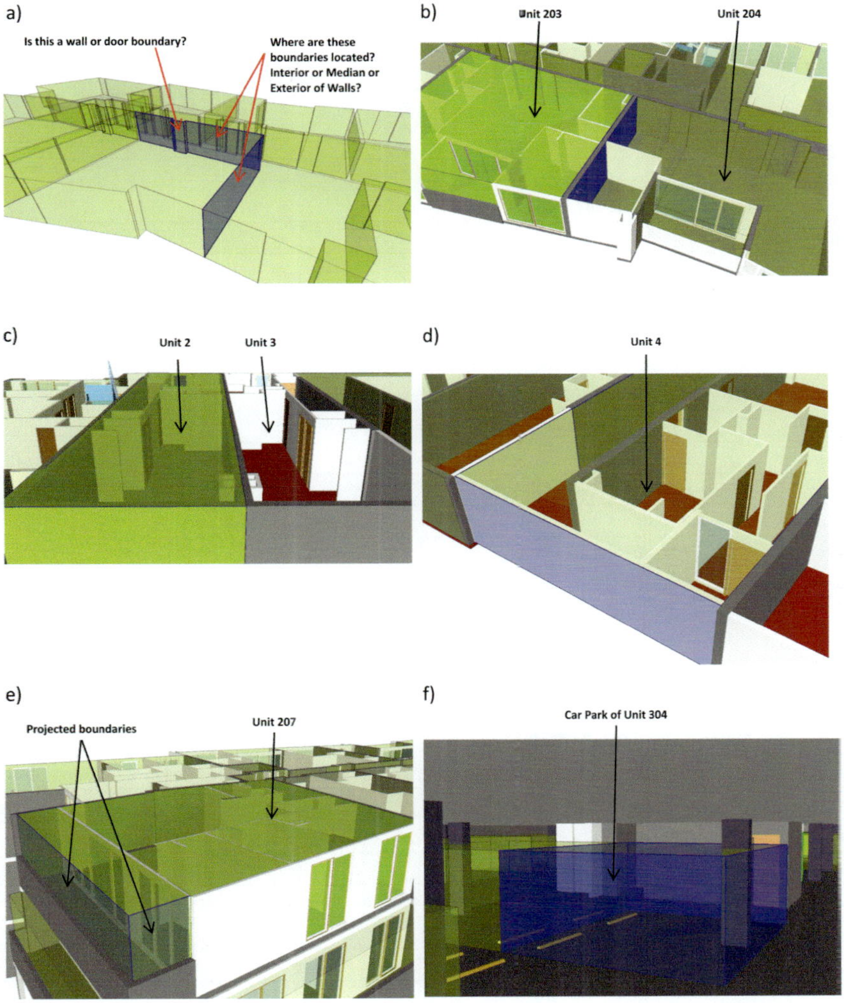

FIGURE 5.11
(a) A purely legal representation in the BIM model; (b) Examples of interior building boundaries; (c) An example of a median building boundary; (d) An example of an exterior building boundary; (e) A projected boundary example; (f) A fixed boundary example.

area. This is more clearly communicated in a BIM environment, which can provide an integrated and interactive view of all legal spaces.

5.5.4 Legal Documents and Land Administration Actors

Information describing legal documents and land administration actors was also defined and linked to the prototype BIM models. The IfcDocumentInformation entity was instantiated to manage metadata for each legal document. For example, Figure 5.12a shows metadata for the a)

Title Document Information	
Property	**Value**
Confidentiality	Only Land Victoria has access to the document
Creation Time	2013-03-20
Description	Volume No. 10546, Folio No. 725, Security No. 6134
Document Owner	Land Victoria
Editors	Mr. Chris Mcray, Mr. David Clarke
Electronic Format	PDF
Intended Use	For registering the ownership of Lot 303
Last Revision Time	2013-0-28
Location	https://www.landata.vic.gov.au/TitleDocument/Title10564.pdf
Name	Title for Lot 303
Purpose	To provide official record of proprietorship of Lot 303
Revision	Final Revision
Scope	No Scope is defined
Status	FINAL
Valid From	2013-03-28
Valid Until	2113-03-28

b)

Approval attributes	
Property	**Value**
ApprovedResources	https://data.melbourne.vic.gov.au/PlanningPermits/PP2031A.PDF
Description	This is permission for developing and subdividing land allocated in 743 Albert Street
Giving Approval	Melbourne City Council
Identifier	PP2031A
Name	Approval of planning permit
Qualifier	Utility services and access ways must be provided for all lots
Requesting Approval	Mr. Tom Champion L.S.
Status	Approved
Time Of Approval	2012-01-03

c)

IfcRelAssignsToActor	
Property	**Value**
Acting Role	Manage
Actor Type	IfcPersonAndOrganization
Related Objects	Common Property No. 1
Relating Actor	Mr. Chris Williamson - Knight Alliance

FIGURE 5.12
(a) An example of metadata attributes about legal documents; (b) An example of approval attributes assigned to legal documents; and (c) Assignment of the manager of an Owners Corporation to the common property.

title document which is certified to the owner of LOT-303. These attributes can be viewed through the attribute manager of SMV. Similarly, information describing other legal documents, such as mortgages and covenants, can be managed and linked to the BIM model. In addition to metadata, the IfcApproval entity can be also instantiated to model the approval process for various legal documents such as planning permits, statements of compliance, certifications, and engineering plans. For instance, the approval attributes of the planning permit for subdividing development C is extracted from the attribute manager of SMV and shown in Figure 5.12b. Finally, different types of land administration actors were assigned to their corresponding legal interests. For instance, Figure 5.7c shows that the manager of the Owners Corporation was assigned to the common property through instantiating the IfcRelAssignsToActor entity.

5.5.5 Administrative Information

Figure 5.13 shows an overall spatial extent of the implemented prototype BIM model for development C. As mentioned in Chapter 4, plan information and site attributes are the two main types of administrative information used in urban land administration. The plan information, which is defined as a property set of the IfcProject entity, for development C is represented

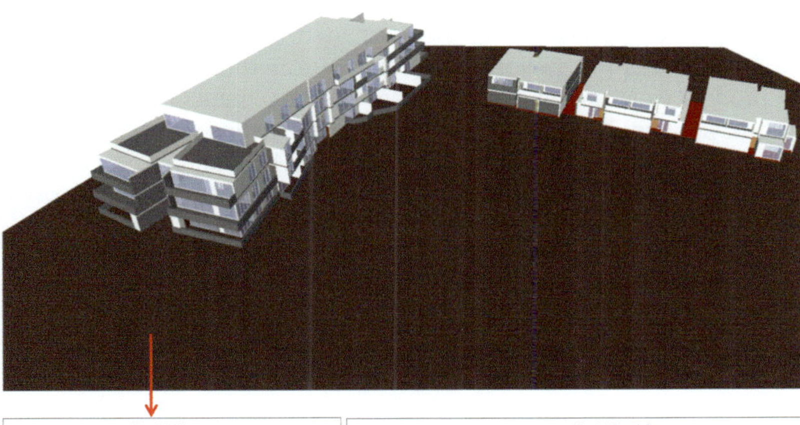

Site Attributes	
Property	**Value**
RefElevation	24.6 m
RefLatitude	-37 47 7.11
RefLongitude	145 1 24.13
SiteAddress	9-23 Rex Avenue, Alphington VIC 3078

Pset_PlanInfo	
Property	**Value**
Administrative Date	2012-05-03
Head Of Power	Subdivision Act 1988 and Subdivision (Registrar) Regulation 2011
Lots No	58
Plan No	PS644630X
Stage No	2
Status	Application for certification
Survey Format	Volumetric

FIGURE 5.13
Example of administrative information managed within BIM models.

in Figure 5.13. The plan information includes important attributes such as the administrative date, number of lots, plan number, stage number, status of the application, and format of the survey. In addition to these attributes, the attributes describing the geospatial location of the development are also important for administrative purposes. The geospatial location attributes, namely latitude, longitude, height, and the site address, can be defined by the IfcSite entity. As an example, these attributes for development C are represented in Figure 5.13.

5.6 Concluding Remarks

In this chapter, to demonstrate the feasibility of the BIM-driven approach for urban land administration, prototype BIM models were implemented for simple and complex developments located in an urban built environment. These implemented prototype models provide the practicality of recording, managing, and visualizing legal dimensions of urban built environments within the 3D digital data environment of BIM. The chapter considered two approaches for constructing BIM models: 3D design and 3D surveying. Most existing BIM models are constructed during the design phase of a building, and they are not updated after the construction phase of the building. Therefore, in this chapter, various 3D surveying methods for creating as-built BIM models were also presented. Legal arrangements of buildings in the real world may be different to legal boundaries delineated based on the design model of buildings. Therefore, as-built BIM models are needed to check the conformance of a BIM model with its building in the real world to ensure the integrity of the BIM model.

References

Bassier, M., R. Klein, B. Van Genechten, and M. Vergauwen. 2018. "Ifcwall Reconstruction From Unstructured Point Clouds." *ISPRS Annals of Photogrammetry, Remote Sensing and Spatial Information Sciences* IV–2:33–39. doi:10.5194/isprs-annals-IV-2-33-2018.

Brenner, Claus. 2005. "Building Reconstruction from Images and Laser Scanning." *International Journal of Applied Earth Observation and Geoinformation* 6 (3–4):187–198.

Chen, Liang Chien and Tee Ann Teo. 2004. "Fusion of LIDAR Data and Optical Imagery for Building Modeling." *International Archives of Photogrammetry and Remote Sensing* 35.

Chen, Liang Chien, Tee Ann Teo, Jiann Yeou Rau, Jin King Liu, and Wei Chen Hsu. 2005. "Building Reconstruction from LIDAR Data and Aerial Imagery." In: *International Geoscience and Remote Sensing Symposium (IGARSS)*, 29 July 2005, Seoul, South Korea.

Habib, A., M. Ghanma, and M. Tait. 2004. "Integration of Lidar and Photogrammetry for Close Range Applications." *XXth ISPRS Congress: Geo-Imagery Bridging Continents*, 12–23 July 2004, Istanbul, Turkey.

Ham, Youngjib, Kevin K. Han, Jacob J. Lin, and Mani Golparvar-Fard. 2016. "Visual Monitoring of Civil Infrastructure Systems via Camera-Equipped Unmanned Aerial Vehicles (UAVs): A Review of Related Works." *Visualization in Engineering* 4 (1).

Hamledari, Hesam, Brenda McCabe, Shakiba Davari, and Arash Shahi. 2017. "Automated Schedule and Progress Updating of IFC-Based 4D BIMs." *Journal of Computing in Civil Engineering* 31 (4).

Jazayeri, Ida, Abbas Rajabifard, and Mohsen Kalantari. 2014. "A Geometric and Semantic Evaluation of 3D Data Sourcing Methods for Land and Property Information." *Land Use Policy* 36:219–230.

Keqi, Zhang, Yan Jianhua, and Shu Ching Chen. 2009. "Automatic 3D Building Reconstruction from Airborne LiDAR Measurements." *2009 Joint Urban Remote Sensing Event*, 20–22 May 2009, Paris, France. doi:10.1109/URS.2009.5137548.

Kurul, Esra, Henry Abanda, Joseph H. M. Tah, and Franco Cheung. 2013. "Rethinking the Build Process for BIM Adoption." In: *CIB World Building Congress Construction and Society*, 5–9 May 2013, Brisbane, Australia.

Nguyen, Anh and Bac Le. 2013. "3D Point Cloud Segmentation: A Survey." In: *Proceedings of the IEEE Conference on Robotics, Automation and Mechatronics, RAM*, 12–15 November 2013, Manila, the Philippines.

Oude Elberink, Sander and George Vosselman. 2011. "Quality Analysis on 3D Building Models Reconstructed from Airborne Laser Scanning Data." *ISPRS Journal of Photogrammetry and Remote Sensing* 66 (2):157–165.

Remondino, Fabio and Sabry El-Hakim. 2006. "Image-Based 3D Modelling: A Review." *The Photogrammetric Record* 21 (115):269–291.

Rottensteiner, Franz. 2003. "Automatic Generation of High-Quality Building Models from Lidar Data." *IEEE Computer Graphics and Applications* 23 (6):42–50.

Schnabel, R., Raoul Wessel, R. Wahl, and Reinhard Klein. 2008. "Shape Recognition in 3d Point-Clouds." In: *Proceedings of the 16th International Conference in Central Europe on Computer Graphics, Visualization and Computer Vision 2008*, 4–7 February 2008, Plzeň, Czech Republic.

Shan, J. and C. Toth. 2010. "Book Review: Topographic Laser Ranging and Scanning: Principles and Processing." *International Journal of Remote Sensing* 31(12): 3333–3334.

Sohn, Gunho and Ian Dowman. 2007. "Data Fusion of High-Resolution Satellite Imagery and LiDAR Data for Automatic Building Extraction." *ISPRS Journal of Photogrammetry and Remote Sensing* 62 (1):43–63.

Vo, Anh Vu, Linh Truong-Hong, Debra F. Laefer, and Michela Bertolotto. 2015. "Octree-Based Region Growing for Point Cloud Segmentation." *ISPRS Journal of Photogrammetry and Remote Sensing* 104:88–100.

Vosselman, George, Maximilian Coenen, and Franz Rottensteiner. 2017. "Contextual Segment-Based Classification of Airborne Laser Scanner Data." *ISPRS Journal of Photogrammetry and Remote Sensing* 128:354–371.

Vosselman, George and Er Dijkman. 2001. "3D Building Model Reconstruction from Point Clouds and Ground Plans." *International Archives of Photogrammetry, Remote Sensing and Spatial Information Sciences* XXXIV (3/W4):37–44.

Wang, Zheng and Tony Schenk. 2000. "Building Extraction and Reconstruction from LiDAR Data." *International Archives of Photogrammetry and Remote Sensing* XXXIII (B3):958–964.

Wolf, Daniel, Johann Prankl, and Markus Vincze. 2015. "Fast Semantic Segmentation of 3D Point Clouds Using a Dense CRF with Learned Parameters." In: *Proceedings – IEEE International Conference on Robotics and Automation*, 26–30 May 2015, Seattle, WA. doi:10.1109/ICRA.2015.7139875.

6

Impacts of BIM on Urban Land Administration

6.1 Introduction

BIM is not a new paradigm in the AEC industry. Indeed, over the last decade, BIM has been in action to enable a 3D digital approach for designing and constructing multi-story buildings. Within the AEC industry, BIM has had a significant impact on resolving communication and management issues associated with 2D drawings and even 3D CAD models (Arayici et al. 2011; Mihindu and Arayici 2008). Unlike the AEC industry, the land administration industry has not adopted a 3D digital practice for modeling and mapping land and property information in multi-story buildings. As mentioned in Chapter 2, land administration practices in most jurisdictions still mainly rely on 2D drawings to represent the legal extent of stratified properties which exist predominantly in an urban built environment.

BIM can be a good source of 3D spatial information to enable digital urban land administration. BIM creates a new opportunity for the land administration industry to enhance the communication and management of legal ownership for properties in multi-story buildings with a wider community of stakeholders (Atazadeh, Rajabifard, and Kalantari 2017). With more and more AEC practices using BIM for managing the building lifecycle, there will be more sources of 3D digital data which can potentially be used for security of property tenure in multi-story buildings. This will change the land administration industry in the immediate future and long term.

In the short term, the land administration industry can benefit from the 3D digital data that is provided in the BIM environment. A BIM-driven approach can transform current practices, which depend on 2D analogue plans of subdivision, to a 3D digital environment. In the long term, BIM will shift land administration from a practice which solely ensures the security of property ownership to a practice that enables the use of 3D digital land and property information for not only managing ownership arrangements in complex urban environments, but also with the potential to be leveraged for broader urban applications. One main application outside the domain of land administration is encoding the legal extent of properties in a BIM environment for facility management.

Land administration systems have been using 2D-based architectural plans as an input to create survey plans for a long time. It is an established practice where details on an architectural plan are generalized to create and depict property boundaries. If BIM replaces architectural plans in the design stage, the existing practice is maintained by using 3D digital data, instead of architectural plans, to create and depict property boundaries. In other words, land administration practices can continue to use 3D models without major modification in their fundamental processes. In this change, what will be helpful for the end users is a better understanding of property boundaries; for the land surveyor, more flexibility and options for communication of the RRR; and for the responsible authorities, such as land registries and local governments, it will be easier to manage and examine complex RRR information for properties in multi-story buildings.

As an integrated and multi-disciplinary approach to managing buildings, BIM will also push the land administration industry to move away from its focus on plan creation to communicating legal arrangements of stratified properties in a more visual way and adopting 3D digital data management methods.

In this chapter, the benefits of a BIM-driven approach for urban land administration will be described. In addition, the chapter highlights the advantages of a BIM-driven urban land administration for managing facilities, particularly communal assets, in multi-story buildings. Finally, this chapter describes major technical and institutional challenges that could be faced in the adoption of a BIM-driven approach in land administration practices.

6.2 Benefits of BIM for Urban Land Administration Practices

The prototype BIM models presented in Chapter 5 demonstrated the clear advantages that exist in overcoming communication and management challenges. Visualizing legal interests and boundaries in a 3D digital environment is easy to understand and supports the community's understanding of who owns what, and where. In particular, the BIM environment provides the ability to model the spatial structure of legal arrangements and connective boundary relationships which supports the practice of using a building's physical structure to define the legal ownership of properties. The potential benefits of the BIM environment for urban land administration can be summarized as:

- Communication benefits: The interactive, integrated, and 3D digital data environment of BIM would facilitate the communication of legal information with a wider community of stakeholders with a limited background in a property subdivision profession. This mainly includes improving the understanding of legal boundaries and distinguishing boundary types from each other within the BIM environment.

- Management benefits: In complex and multi-story developments, spatial and physical elements constituting each legal interest are typically dispersed in various parts of these developments. The BIM environment would facilitate management of legal interests, since it has the capability to arrange spatial and physical elements into the spatial zone of each legal interest by using spatial relationships. This would also make it easier to find and query different parts of each legal interest.

The benefits of a BIM-driven approach in terms of modeling legal interests and boundaries are described in further detail below.

6.2.1 Legal Interests

In current practice, the spatial extent of legal interests is typically delineated in various floorplans and cross-section diagrams. Interest holders often find it difficult to identify and locate relevant information pertaining to a particular legal interest within multiple pages of 2D plans (see Figure 6.1a). BIM models usually store building data in a hierarchical data structure (building, building story, spaces, and so on) and users can leverage this hierarchy to navigate and locate building elements. Finding legal spaces associated with a specific property might be difficult for the interest holder of that property inside large and tall buildings due to complex physical structures and the large number of legal spaces defined inside each building story. However, the proposed IFC extension presented in Chapter 4 provides the BIM environment with the ability to arrange various legal spaces and physical elements into a non-hierarchical spatial structure, such as a spatial zone or zone, using semantic relationships. For instance, the spatial arrangement of a multi-part lot can be defined by arranging the legal spaces defining its constituent parts (apartment unit, carpark, and storage area) into a spatial zone using spatial relationships (see Figure 6.1b).

As mentioned in Chapter 2, building elements are sometimes used in defining the whole spatial structure of legal interests. For instance, the building elements, which are part of common property areas, are not represented in current 2D-based practices (see Figure 6.2a); they are simply described using notations. 2D subdivision plans provide an abstract representation of the physical structures in 2D plans. This abstraction may not provide a clear image of the spatial extent of common properties for the responsible owners corporations. Therefore, during the building management phase,

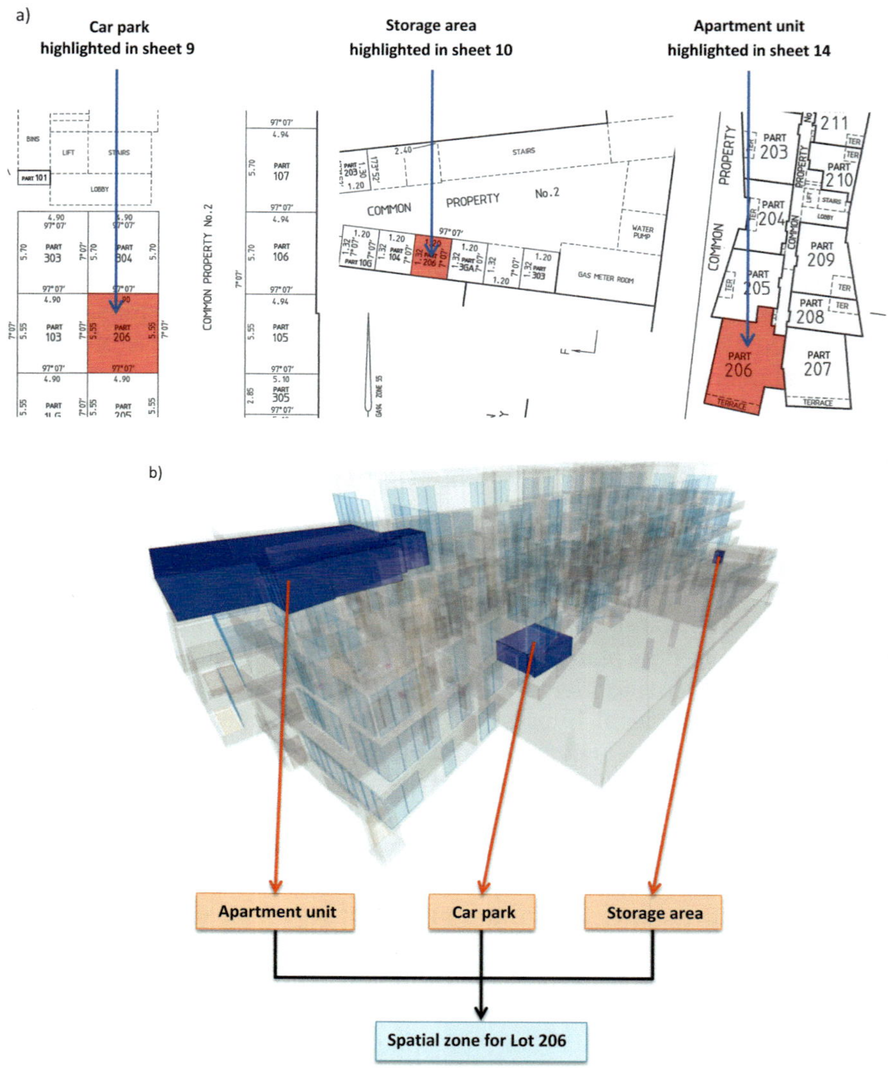

FIGURE 6.1
An example of a volumetric lot represented in (a) subdivision plan and (b) the BIM environment.

owners corporations often face significant challenges in identifying which part of a development belongs to the common property in spatially complex situations.

The full extent of the legal spaces defining common properties can be visually communicated in the BIM environment. In addition, the building elements deemed to be part of a common property are not abstracted in the BIM environment making it easier to identify those building elements belonging to the common property. Therefore, as shown in Figure 6.2b, the

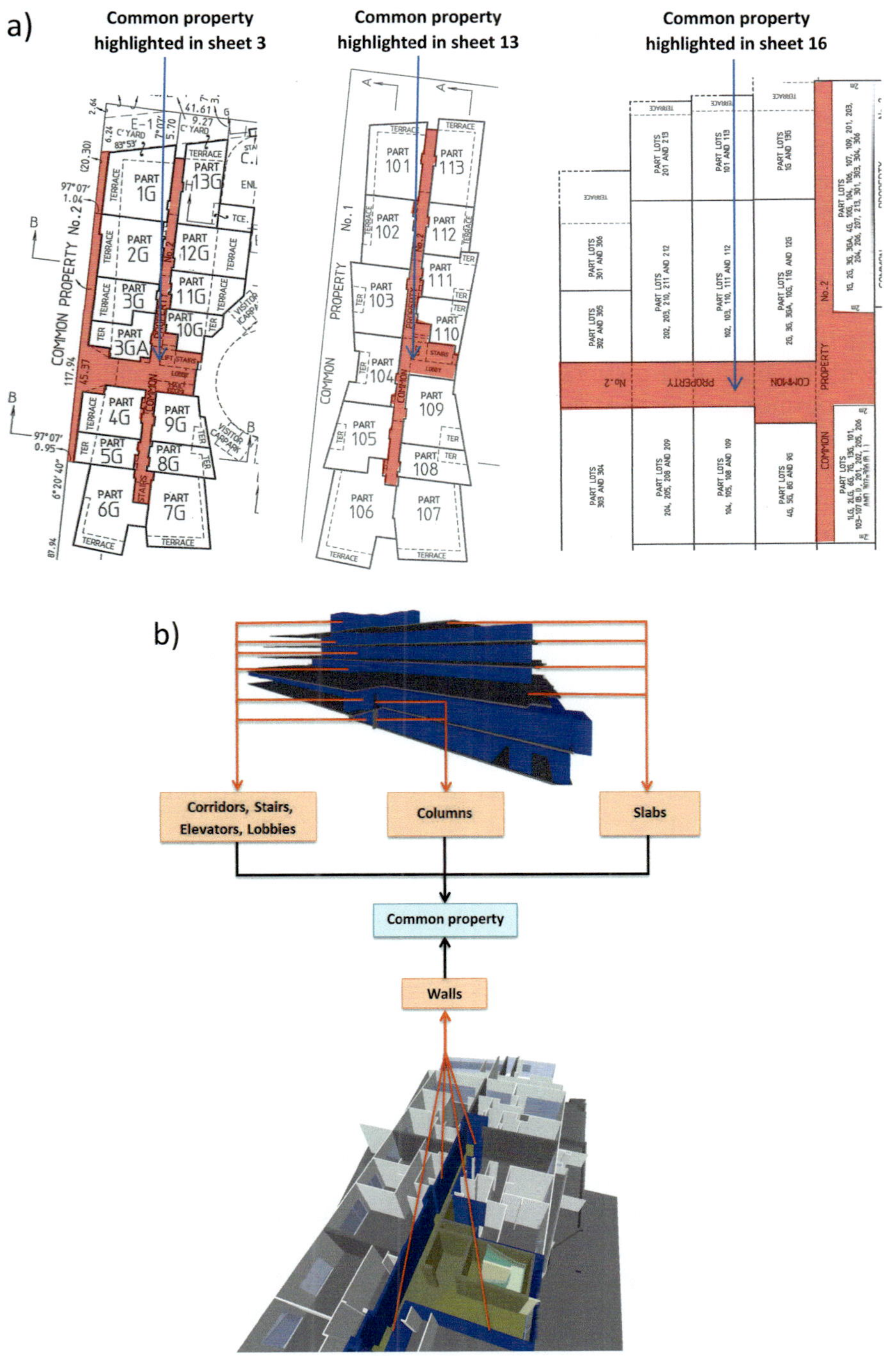

FIGURE 6.2
Part of the common property represented in (a) a subdivision plan; and (b) the BIM environment.

representation of building elements as part of the common property spatial zone in a BIM environment could help owners corporations visually perceive the building elements for which they have the responsibility of management and maintenance.

From a technical point of view, a BIM-driven approach would also help to manage multiple common properties defined inside complex urban developments in a more transparent and communicable way (Atazadeh et al. 2018). To define a fair approach for managing maintenance costs, unlimited and limited owners corporations are typically created in complex urban developments. Unlimited owners corporations have unlimited functions, powers, and responsibilities throughout the whole development. Each limited owners corporation typically serves a specific part of the common property area, which is used by a group of lot owners residing within the development. During the lifecycle of common properties in a multi-story development, the function and tasks of a limited owners corporation are ancillary to the overarching unlimited owners corporation. In 2D-based practices, a clear understanding of the spatial extent of both unlimited and limited common property areas could be a challenging task for managers of owners corporations. A BIM-based approach for managing common properties would potentially facilitate collaboration among multiple owners corporations during the building lifecycle by providing a more transparent, integrated, and 3D digital representation of the complex spatial structure of common properties. Therefore, the BIM environment could also have a significant impact on the interactions among various owners corporations and the governance of common properties in a multi-story development.

Incorporating legal information into the BIM environment also provides benefits in managing buildings, since legal information plays an important role throughout the lifecycle of buildings. For instance, legally defined entitlements and liabilities of properties are fundamental to building and property management where: lot entitlement is a number expressing the extent of each private property owner's interest in any common property; and lot liability is a number expressing the proportion of the administrative and general expenses that each private property owner is obliged to pay. In the ongoing management of common properties in a building, entitlements and liabilities provide the legal basis for dealing with disputes and dissatisfaction. Therefore, BIM models enriched with legal information would increase the functionality of these models in terms of managing the lifecycle of buildings.

Another significant benefit of a BIM-based solution is the ability to support 3D spatial queries, volumetric computations, and 3D spatial analyses required in urban land administration. Currently, it is difficult to measure and compute volume of legal interests using 2D plans, since a fragmented mapping of volumetric legal spaces into cross-section diagrams and floorplans does not provide an efficient mechanism to calculate volumetric quantities. In contrast, the BIM environment supports a wide variety of solid models, such as B-rep, sweep, and CSG, for modeling the 3D shape of objects. These solid models provide the BIM environment with the ability to not only create and represent valid volumetric RRR spaces with various spatial complexities, but to also carry out 3D computations such as volume measurement. In addition, 3D digital data in the BIM environment can be reasoned using 3D analyses required in urban land administration, such as the 3D visibility analysis of properties useful in valuation and taxation purposes.

Overall, using a BIM-driven approach, current urban land administration practices could potentially progress to a 3D digital, collaborative, and dynamic data environment. This environment would facilitate communication, management, and analysis of complex legal interests in multi-story buildings, which would subsequently enhance the engagement of the land administration industry with a broader community of stakeholders involved in the building development process.

6.2.2 Legal Boundaries

To describe the benefits of a BIM-driven approach for urban land administration, in this section, various legal boundaries inside the prototype BIM models are compared with their counterparts delineated in the current 2D-based practice. The comparison outlined below describes how the BIM environment would enhance the communication of legal boundaries by improving on the visual representation.

- Interior structural boundaries: Figure 6.3a represents an apartment unit (unit 203) and the common property area within the second level of the apartment building, in which an interior boundary for unit 203 is highlighted in both the plan version and the BIM environment. For a user with no specialist background in understanding subdivision plans, it is hard to interpret the location of the boundary in this case. In contrast, inside the BIM environment, it can be intuitively perceived that the boundaries of the apartment units are delineated by the internal surfaces of the wall. This would communicate that the owner of the apartment unit does not hold any rights to the wall, which consequently belongs to the common property.
- Median structural boundaries: Some townhouses are represented in Figure 6.3b, which shows two abutting townhouses (units 2 and 3) and the position of their common median boundary in both the subdivision plan and BIM environment. A land surveyor might define a median boundary using an "M" annotation alongside the boundary line. However, the owners of these units may have difficulty in

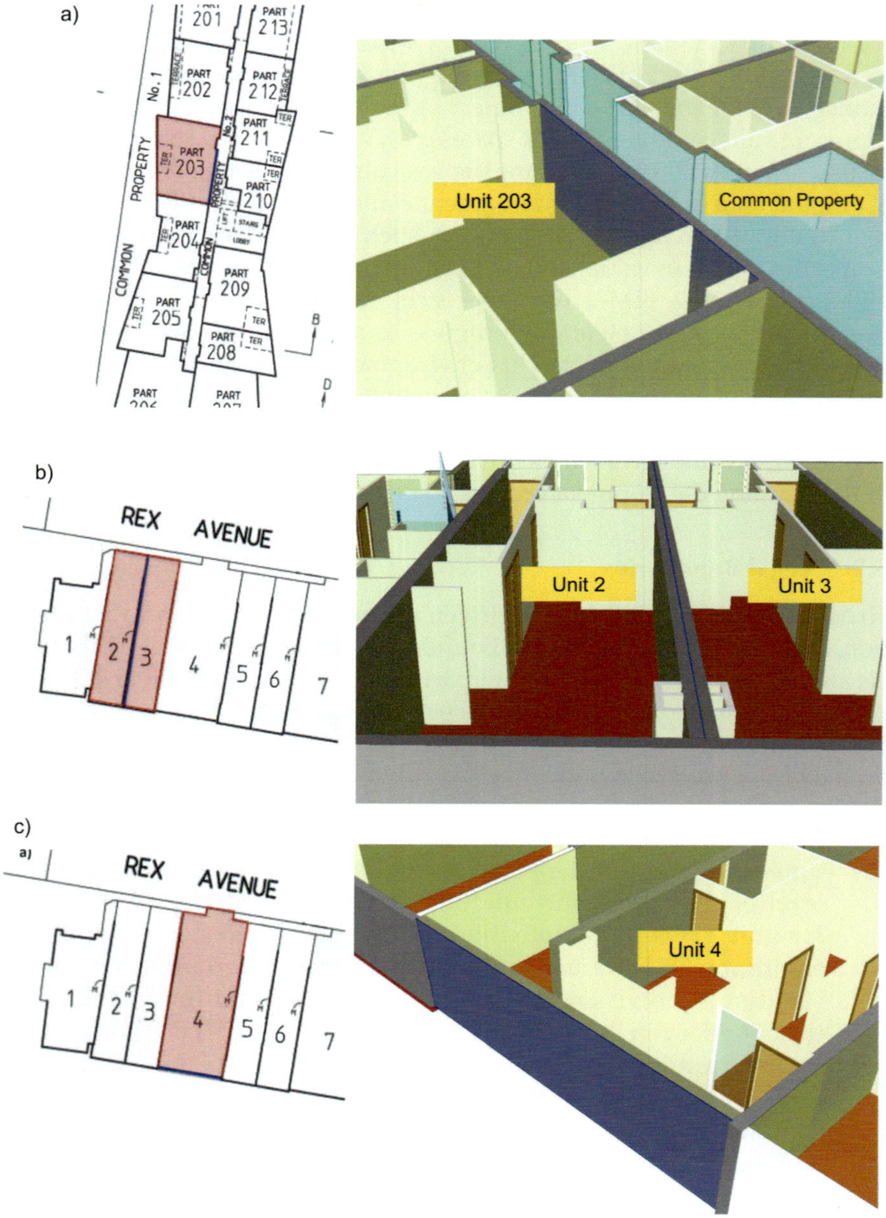

FIGURE 6.3
Building boundaries represented in subdivision plan and the BIM environment: (a) An interior boundary; (b) A median boundary; (c) An exterior boundary.

reading and understanding this boundary. When represented in a BIM environment, the spatial location of this boundary is communicated unambiguously to the owners. Each owner can clearly understand that ownership of the wall shared among them.

- Exterior structural boundaries: As illustrated in Figure 6.3c, a typical example of an external boundary for unit 4 is highlighted in both the subdivision plan and the BIM environment. In the notation section of the subdivision plan, it is implicitly indicated that this boundary is an external one. However, the owner of unit 4 may not accurately interpret its location. Mapping this boundary into a BIM environment could be more congruous with the visual perception of non-specialists. This would easily communicate that the building element associated with the boundary (in this case, the wall) completely belongs to unit 4.
- Projected boundaries: A BIM environment can represent projected boundaries automatically by extruding the physical boundary surfaces in both the vertical and horizontal directions (see Figure 6.4a). In subdivision plans, these can be misinterpreted similarly to structural boundaries, since their location is dependent on the position of their corresponding structural boundaries. However, modeling in a BIM environment can facilitate communication of these boundaries in the same way as structural boundaries.
- Fixed boundaries: Figure 6.4b shows how fixed boundaries associated with a car park area can be represented in both the subdivision plans and the BIM environment. Interpretation of fixed boundaries in subdivision plans can also be difficult for owners. For instance, as illustrated in Figure 6.4b, azimuth values annotated alongside boundaries of a carpark belonging to unit 1G may seem strange for the owner, and he/she might get confused in terms of interpreting those values. Another issue is that owners may only look at the floorplan of their car park area and think that the legal space of their car park is up to the ceiling. However, in cross-section diagrams it is sometimes indicated that the vertical extension of the car park is only up to two meters. The remaining space from a height of two meters to the ceiling is part of the common property area. This is more clearly communicated in a BIM environment, which can provide an integrated and interactive view of all legal spaces.

Modeling 3D digital legal boundaries within the BIM environment could also support the facility management industry in repairing, operating, and maintaining assets within multistory buildings. The position of boundaries within building elements would assist in determining whether the liability for repairs rests with the private owner, the owners corporation, or a combined responsibility of both parties. For example, if the ownership

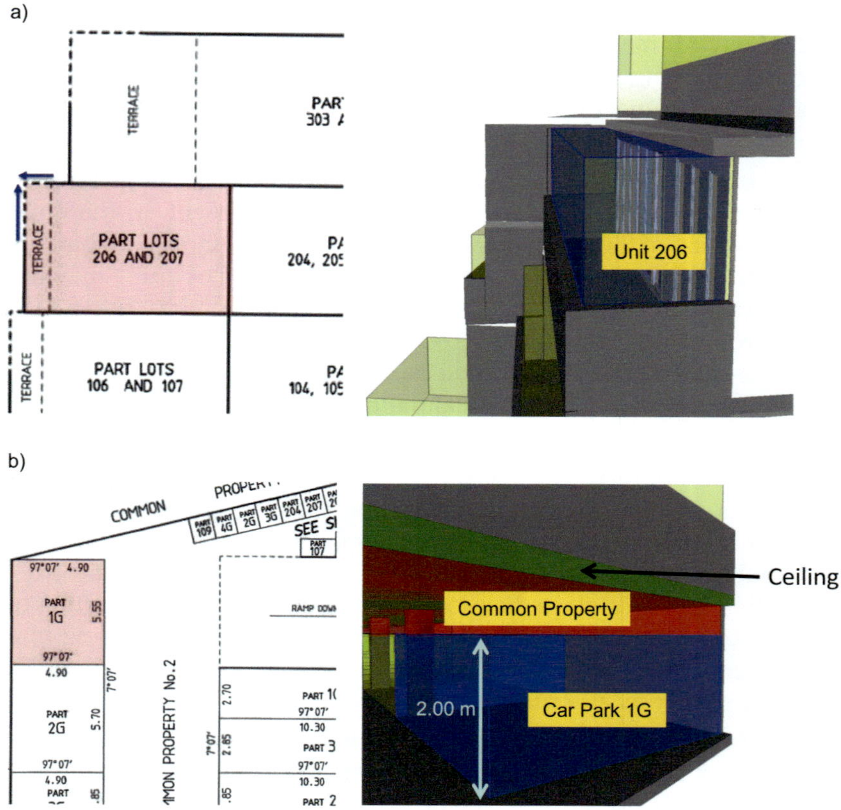

FIGURE 6.4
(a) Projected boundary represented in subdivision plan and the BIM environment; (b) Fixed boundary represented in subdivision plan and the BIM environment.

boundary is associated with the interior faces of the building elements, this would indicate that these building elements, as well as any asset inside them, such as conduits and pipes, are part of the common property. In this case, the owners corporation would have responsibility for maintaining those building elements and assets. However, in the case of median boundaries between two private properties, the ownership of corresponding building elements is halved between owners. For example, if there was damage throughout the whole building element, the BIM environment can visually support the fact that both owners would have an equitable and combined responsibility for repairing the damage. For exterior boundaries, the BIM model would show that the building elements are completely enclosed within a property. This would indicate that the owner of that property has full rights of use and, therefore, responsibility for repairing any damage to the assets within them.

6.3 Challenges of BIM-Based Urban Land Administration

Although a BIM-based approach provides significant benefits for urban land administration, there are some barriers in current procedures to fully implementing and adopting BIM. These obstacles are classified into two categories: technical and institutional challenges.

6.3.1 Technical Challenges

To fully implement the BIM environment aligned with current urban land administration practices, there are a number of technical obstacles that need to be overcome. These challenges arise when recording, managing, and communicating 3D digital information. One technical issue most BIM models face is a discrepancy between the design and construction phases of a building, where the model may not precisely represent the reality of boundaries in the real world. Therefore, verification of the design BIM model is required by land surveyors to check its conformance with the as-constructed model in reality.

Another technical obstacle is associated with information redundancy in architectural BIM models. Not all of the physical building elements are used for modeling the spatial extent and boundaries of legal zones. For example, walls and furniture, such as desks and chairs, located within apartment units are not required for managing legal spatial zones. To overcome this issue, BIM models should be generalized by removing such inessential physical elements. This generalization would also reduce the data volume of the BIM model, which would, in turn, enhance the speed when visualizing and querying the model.

These issues outlined are two examples of major technical challenges that may occur when a BIM-driven approach is adopted for urban land administration. The technical challenges can be resolved by adopting appropriate techniques; however, the more difficult and fundamental challenges are the institutional challenges that need to be addressed on the path towards a BIM-enabled urban land administration.

6.3.2 Institutional Challenges

The major barriers for using the BIM and IFC standard are most likely to be the institutional challenges. The current institutional arrangements for managing legal interests in urban built environments are highly entrenched in using 2D subdivision plans (Ho and Rajabifard 2016). Ho, Rajabifard, and Kalantari (2015) identified a range of "invisible" constraints, namely, regulatory, normative, and cultural–cognitive, which are impeding the move towards 3Denabled urban land administration practices, and would arguably apply to the adoption of the BIM environment in existing urban land administration procedures (see Figure 6.5).

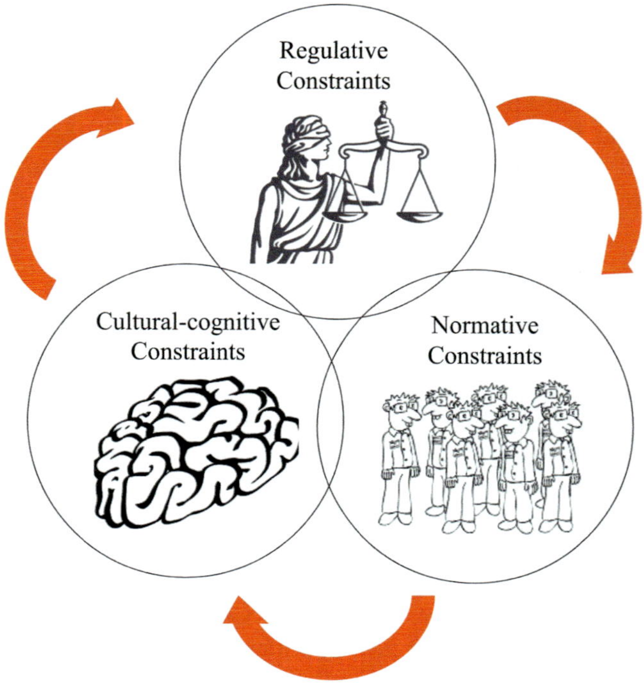

FIGURE 6.5
Institutional challenges on the path towards a BIM-enabled urban land administration.

This section applies institutional theory to identify plausible ways in which current stakeholders are being constrained to the current ways of working by the pressure exerted through the regulatory, normative and cultural–cognitive elements within the current institutional environment. The constraints are framed around the core institutional concept of legitimacy: constraints are therefore those elements that support or sustain legitimacy around the current 2D paradigm of representing land and property information for land administration purposes. These constraints are elaborated in further detail in the following subsection.

6.3.2.1 Regulative Elements as Constraints

The perceived dominance and rigidity of the legislative framework could be interpreted as a constraint on change from 2D-based practices to the 3D BIM environment in several ways. Although the current information processes and products are not keeping pace with the requirements associated with vertical living and the stratified use of space in urban areas, opportunities for change are not identified readily (Ho 2014). This could be mainly attributed to key pieces of the legislative framework having been in operation for decades. For instance, in Victoria, Australia, the current regulatory framework for development was established in the late 1980s, and its institutional force is evident in a widespread acceptance of the logic of the requirements prescribed by the Subdivision Act 1988—a highly routinised process which till now, has been simply taken for granted. Such an extensive duration also implies ample time for the industry to become familiar with its requirements and idiosyncrasies, which support its dominance as an institutional driver at both a conscious and an unconscious level.

This also indicates a significant degree of path dependency in the system, likely due to the legislative framework producing learning effects, where skills and knowledge gained in terms of subdivision and registration have efficiencies attuned towards broad acre, lower density subdivisions that are the majority. This suggests the presence of a coercive pressure to abide by current 2D-based ways of working, i.e. the perception is that there is no choice but to do so. This pressure is not insignificant, especially as it does not seem that the law actually explicitly rules out the use of 3D digital information in the BIM environment. In this instance, there is also significant normative and mimetic pressure inherent in the institutional environment, which results in an unconscious meaning being attached to terms such as "plans" and "documents." These are traditionally associated with 2D-based information but appear to be construed as legitimizing 2D methods. Alternatively, this could also be interpreted as a lack of legitimacy around 3D digital innovation. Importantly, such a lack of explicit support for BIM-enabled urban land administration is interpreted as a restrictive factor for innovation.

The current structure also does not provide any enablers and continues to constrain them to current (2D-based) behavior—a clear emergence of an institutional "gap." There is evidence of attempts made to plug these gaps, such as the development of new regulations, but these are still wholly aligned with current 2D-based ways of thinking and acting. The current framework is also oriented around registration, and therefore participants associate the role of the key actor with the land registry.

6.3.2.2 Normative Elements as Constraints

A move towards a BIM-based urban land administration demands the use of 3D digital land and property information to fully leverage its benefits in improved querying abilities and reuse and representation of this information. This requires wholesale changes to how 3D digital land and property data is collected, drawn, exchanged, and even registered—a fundamental shift in the norms and values relevant to collaboration, visualization, and analytical processes (Ho 2014). Normative constraints are interpreted to relate to four main aspects:

- The land administration profession being more oriented towards regulative elements
- Traditional use of 2D methods for representing building information

- Content of 2D subdivision plans intended only for registration and governed by the relevant legislation
- Lack of awareness of how poor building information is a macro-level issue

First, in terms of the perceived dominance of regulative elements in the institutional environment, the significance associated with these elements could be interpreted to be a function of the land administration profession itself. Land surveyors need to undertake specific licensing requirements to practice as cadastral surveyors, and this forms the basis for professional legitimacy. Their work is often audited and checked by the responsible authorities, and any non-adherence could result in the loss of their license and professional sanctions, thus fostering a strong culture of compliancy. Their professional perspective could therefore be interpreted to be more oriented than usual towards regulative elements, given that the nature of their work is typically in accordance with legal specifications.

Second, the longstanding tradition of representing building information in 2D formats could be interpreted as a significant normative constraint. Drafting conventions encode knowledge, with each of the design professions having its own frame of reference: "interpretive schema that simplifies and condense the "world out there" by punctuating and encoding objects, situations, events, experiences, and sequences of action" (Snow and Benford 1992, 137). These conventions are reinforced and reproduced through education and professional standards, and are expected of these professions. There is little doubt that although BIM-based methods of representing building information are gaining ground, and social expectations around 3D BIM models are growing, it has yet to achieve a level of acceptability within the regulated environment of land administration that is implicitly associated with 2D methods. Although current drafting standards are being challenged by the requirements of complex structures, there is still little movement in shifting to a BIM-based paradigm.

The growing variance in professional skills and abilities to tackle these complex structures can also be interpreted to be a function of the small pool of surveyors who currently engage with these projects, and the long tail of surveyors who dabble in them from time to time. This signals gaps in normative frameworks such as procedures and approaches, where skills are less developed in the context of multi-story structures.

Third, the plans and the legal information represented within them were originally intended for registration purposes only. This original intention and narrow focus have resulted in established "silos" of practice and governance. A common one is an apparent disconnect between registration and management. The current environment makes little allowance for bringing other stakeholders in—in terms of institutional theory, the organizational field is strongly bounded, and stakeholders with a growing presence, such as strata managers, are struggling to find a way to fit in. This alludes to the lack of information flow to strata managers, and clear issues arising from their co-opting of the prevailing institutional logic around the current practice, represented by 2D plans, and their frustrations with its misfit for management purposes. There is a lack of ownership around the change process from 2D to a 3D digital data environment. This could be interpreted as a constraint, particularly in a regulated industry such as land administration, which would require some explicit expression around the use of BIM-based practices and processes within the legislative framework as one of the key drivers to build legitimacy around change.

6.3.2.3 Cultural—Cognitive Elements as Constraints

Dominant cultural norms can be interpreted to constitute intractable obstacles to reform, and the case study conducted by Ho, Rajabifard, and Kalantari 2015 indicates that there are at least several in existence.

First, the tension expressed by stakeholders with regard to balancing subdivision and registration processes and products that are still effective and efficient for most of the (simpler) multi-story developments. The acceptance of these methods is now culturally embedded, as they have been in existence for several decades.

Second, although strata titling and subdivision practices set out to recognize a different type of tenure, the dominance of simple two-lot subdivisions and broad acre subdivisions are very much still in the mold of traditional freehold dwelling types. Arguably, these practices have been acculturated and are now aligned with the dominant "land" paradigm. To date, current cultural norms around urban land administration practices and processes are a product of the dominant development type in most jurisdictions, which means practices and processes pertaining to complex high-rise subdivisions have very much become a somewhat specialized practice, and are not seen to be a unilateral issue.

Cultural norms can also be framed through social patterns rather than literal instructions. Current regulations may place no restrictions on the medium or format of the information held in the register of land. For instance, in the state of Victoria, the current Subdivision (Registrar's Requirements) Regulations, which mandate the form and content of subdivision plans, does not explicitly reject the use of 3D digital models or 3D representations of the statutory information: Part 2, Section 7.1 states that "The Registrar may accept a plan for lodgment that is … (c) a plan in any other form approved by the Registrar" (State Government of Victoria 2011, 3).

However, within the professions associated with land administration, the tradition of 2D drawings is strong. Even though the plans cost valuable time and money to produce, and despite their detail, are difficult for the community to understand and use, for the professions, 2D is culturally embedded in their training and work experience, it is normatively legitimized by standards and conventions, and in some instances, legally sanctioned. To change this "mindset," mandatory measures are essential.

The intractability of cultural norms can be seen in the limited acceptance and understanding of the legislation, which was developed to better suit the needs of strata communities by expanding the definition of "common property." However, though these changes have the force of legislation behind them, the community's struggle to correlate (and, therefore, understand) the new and old paradigms of common property and their associated rights, restrictions, and responsibilities indicate how cultural norms can present as significant constraints. Finally, given the requirement of personal investment to support technological change, the attitudes of individuals could also be interpreted to be a constraint.

6.4 Concluding Remarks

Adoption and implementation of a BIM-based approach in the land administration domain will enable an accurate, authoritative, and unambiguous recording and representation of 3D digital land and property information, which is fundamental to ensuring security, comprehension, and ongoing commoditization of complex vertical legal arrangements common in urban environments. This chapter highlighted the major impacts of a BIM-driven approach on urban land administration.

As a rich repository of 3D digital data, the BIM environment would potentially facilitate the collaboration and engagement of the land administration industry with other stakeholders, such as facility managers, in governance of complex buildings. In other words, encapsulating legal information into the BIM environment could potentially support those stakeholders who are already using BIM by providing richer information about ownership and RRR which is fundamental to the ongoing management of a multi-story building, particularly its communal facilities. Indeed, BIM can also unlock value in legal information by forming a bridge between the interactive lifecycle and the management of the building.

The ability to achieve BIM-driven urban land administration is inhibited by technical and institutional barriers. The technical issues can be overcome by finding the appropriate technical solutions. However, the existing institutional constraints, which support and reinforce the use of 2D-based land administration practices and processes, are the major challenges faced in adopting the right roadmap for 3D digital urban land administration.

References

Arayici, Y., P. Coates, L. Koskela, M. Kagioglou, C. Usher, and K. O'Reilly. 2011. "Technology Adoption in the BIM Implementation for Lean Architectural Practice." *Automation in Construction* 20 (2):189–195.

Atazadeh, Behnam, Abbas Rajabifard, and Mohsen Kalantari. 2017. "Assessing Performance of Three BIM-Based Views of Buildings for Communication and Management of Vertically Stratified Legal Interests." *ISPRS International Journal of Geo-Information* 6 (7).

Atazadeh, Behnam, Abbas Rajabifard, Mohsen Kalantari, and Jihye Shin. 2018. "A BIM-Driven Approach to Managing Common Properties within Multi-Owned Developments." In: *6th International FIG Workshop on 3D Cadastres*, 201–216, 2–4 October 2018, Delft, the Netherlands. International Federation of Surveyors (FIG).

Ho, Serene. 2014. *Towards 3D-Enabled Urban Land Administration: Invisible Constraints and Strategic Choices*. Ph.D. thesis, The University of Melbourne.

Ho, Serene and Abbas Rajabifard. 2016. "Towards 3D-Enabled Urban Land Administration: Strategic Lessons from the BIM Initiative in Singapore." *Land Use Policy* 57:1–10.

Ho, Serene, Abbas Rajabifard, and Mohsen Kalantari. 2015. "'Invisible' Constraints on 3D Innovation in Land Administration: A Case Study on the City of Melbourne." *Land Use Policy* 42:412–425. Elsevier.

Mihindu, S. and Y. Arayici. 2008. "Digital Construction through BIM Systems Will Drive the Re-Engineering of Construction Business Practices." *Information Visualisation, 2008 International Conference*, 9–11 July 2008, London. IEEE Computer Society.

Snow, David and Robert Benford. 1992. "Master Frames and Cycles of Protest." In: Morris, Aldon and McClurg Mueller, Carol. *Frontiers in Social Movement Theory*, 133–155. New Haven, CT: Yale University Press

State Government of Victoria. 2011. *Subdivision (Registrar's Requirements) Regulations*. Victoria, Australia. Available at: www.austlii.edu.au/cgi-bin/download.cgi/cgi-bin/download.cgi/download/au/legis/vic/consol_reg/srr2011538.pdf.

7

Prospect of BIM in the Land Administration Domain—Technical Aspects

7.1 Introduction

This book has explained the innovative adoption of a widely used 3D digital modeling approach in the architecture, engineering, and construction industry—BIM—for urban land administration and to support the management and representation of legal ownership for properties located inside and around multi-story developments. Compared to current 2D-based practices, the BIM-enabled 3D digital urban land administration provides an enhanced ability to communicate legal boundaries and arrangements of real properties located inside complex urban developments for a broader spectrum of actors and players involved in the urban development process. This book highlighted benefits of a BIM-based approach for urban land administration, which would subsequently have substantial impacts on the governance of urban built environments.

This chapter now focuses on the major directions for the future of BIM-enabled urban land administration. These directions include an integrated management of land administration processes in the BIM environment; extending BIM-enabled land administration for large infrastructure projects; spatial reasoning of legal information in BIM models to enable 3D data analytics capabilities for urban land administration; examining the spatial integrity of BIM models; guidelines for extending the IFC standard for urban land administration in any jurisdiction; linking BIM and property measurement standards; and automating the generalization and simplification of BIM models.

7.2 Integrated Lifecycle Management of Urban Land Administration Processes in BIM

As outlined in Chapter 5, land administration in urban built environments is a complex process. This process requires collaboration among a wide range of land administration actors including land surveyors, developers, local government (city councils), referral authorities, owners, objectors, and owners corporations. During the process, various legal documents are exchanged between land administration actors. In addition, these actors communicate with each other by sharing the required information over the course of each stage of the urban land administration processes. Traditionally, this process was predicated on paper-based drawings for creating and sharing legal documents. This method of storing legal documents is very fragmented and results in silos of legal information during the lifecycle of urban land administration processes, particularly in complex developments. Since the exchanged documents and information are not stored in an integrated data environment, the procedures for lodging, certifying, examining, and registering subdivision plans are slow and costly.

Recently, electronic subdivision planning systems have been adopted to reduce the reliance on paper-based methods of information communication in urban land administration processes. For instance, in Victoria, the Streamlined Planning through Electronic Applications and Referrals (SPEAR) planning system has been developed to compile, lodge, manage, refer, approve, and track urban land administration processes in a digital environment. Although this system provides the capability to manage and track activities of urban land administration processes in an online and digital environment, it still relies on 2D-based methods to record, manage, and communicate legal information.

On the other hand, BIM Level 3 offers a collaborative, 3D digital, intelligent, and semantically enriched data environment to support the integrated management of multidimensional aspects of buildings over their lifecycle. At Level 3, fully open processes that are cloud-based, interconnected, and compliant with the IFC standard are adopted for managing and integrating 3D digital data.

In this book, BIM is mainly adopted from a product perspective for managing spatial arrangements of legal information. The process data elements underpinning urban land administration were also identified. These data elements were modeled by extending the open IFC data model. The outcomes of this extension could be expanded to explore the potential of using a BIM-enabled subdivision planning system to support business processes and information exchanges for the purpose of urban land administration. The design and implementation of such a system can showcase the viability of BIM for managing the entire lifecycle of legal information, from the design phase through to planning and compliance, as well as the registration and management phases in building developments within interconnected, 3D digital and collaborative data environments. Ultimately, this will lead to the next generation of planning and management systems used in urban land administration.

7.3 BIM-Enabled Urban Land Administration in Infrastructure Projects

The urban built environment includes complex infrastructure facilities and utility networks located both underground and aboveground. Typical examples of urban infrastructure facilities include tunnels, bridges, roads, railway tracks, and similar structures adopted in transit systems. Additionally, urban areas include complex utility networks such as telecommunication networks, water supply systems, drainage systems, natural gas pipelines, and electrical powerlines. The legal ownership of infrastructure facilities and utility networks is often held by either private or public authorities.

The spatial arrangements of urban infrastructure are currently often recorded and managed in disparate, isolated, proprietary, and non-interoperable 2D data environments. There are significant challenges associated with recording, managing, and communicating underground and aboveground urban infrastructure data. The data formats are not consistent, which results in an inefficient integration and exchange of spatial information describing urban infrastructure. Due to these inconsistencies, there are often significant costs associated with data duplication in managing urban infrastructure. Currently, the infrastructure industry is slowly shifting from document-centric methods towards 3D digital and intelligent data management technologies.

The major focus of this book was a BIM-driven approach for the 3D digital management of legal ownership for properties located in multi-story buildings, since currently the BIM environment mainly only supports the creation, representation, and management of 3D intelligent models for buildings. However, there is a growing interest in the BIM domain to model the lifecycle of other features, such as infrastructure and landscape features, that exist in an urban built environment. BIM can potentially provide an integrated data environment to support the semantic and spatial aspects of the entire lifecycle of infrastructure assets. Therefore, a BIM-driven approach would lead to the transformation of the infrastructure industry to a 3D digital working space.

In particular, the open IFC standard is being extended for modeling major infrastructure assets such as roads, bridges, tunnels, railways, and similar linear infrastructure developments. The aim of this extension is to provide an open standards-based approach for exchanging 3D digital data over the lifespan of infrastructure projects from design through to operations and maintenance. BuildingSMART defined various major projects in order to extend the IFC standard for the infrastructure domain:

- IFC Alignment: A very important part of an infrastructure project is its alignment. The construction of linear infrastructure assets, such as roads and railways, is heavily dependent on correct alignment.

IFC Alignment is the first project developed recently for extending IFC with infrastructure data elements. This extension of the IFC schema comprises the entities required for a linear 2D and 3D alignment of the spatial location of infrastructure assets. IFC Alignment provides the ability to exchange the alignment data during the lifecycle stages, from planning through to construction to management and maintenance, of infrastructure works. In addition, the alignment information can be connected to other infrastructure information, such as cross-sections, elevations, and the full 3D geometric representation of construction components.

- IFC Infra Overall Architecture: This project provides a shared foundation for the infrastructure-specific extensions of the IFC standard, such as IFC bridge and IFC road. The project defines guidelines and recommendations for the spatial structure of physical components within infrastructure assets, consistent use of horizontal and vertical geodetic datums and map projections, consistent geometric representation of cross-sections and string lines representing edges of linear infrastructure components, using solid models for modeling the volumetric geometry of infrastructure assets, and classification of infrastructure objects.
- IFC extension for each type of infrastructure asset: BuildingSMART chapters in different countries proposed separate projects for modeling various types of infrastructure assets. These projects include IFC Road, IFC Bridge, IFC Tunnel, and IFC Rail. To ensure consistency, these IFC extension developments should be in accordance with IFC Alignment and IFC Infra Overall Architecture projects. In other words, modeling the lifecycle of a particular type of infrastructure asset should be based on the proposed entities defined in IFC Alignment, as well as guidelines suggested in the IFC Infra Overall Architecture.

The above-mentioned projects mainly aim at modeling physical information for infrastructure assets. However, the legal ownership of infrastructure assets plays a fundamental role in planning, constructing, maintaining, and operating these assets for their responsible authorities. Clear and correct definition of legal rights associated with infrastructure assets is important for effective governance of an urban built environment. A reliable and complete representation of legal spaces would avoid unnecessary costs, delays, and disruptions in infrastructure projects. Therefore, a future direction for BIM-driven urban land administration could be harnessing the potential capability of the BIM environment for managing legal arrangements of urban infrastructure assets. A BIM-based digital approach would mitigate the current challenges in subdividing, registering, and managing the legal ownership of infrastructure assets. It can also support better stakeholder involvement and wider community participation, facilitating decision-making processes during the lifecycle of infrastructure projects.

The open BIM environment would be a suitable choice for adopting a BIM-driven approach for the registration and management of volumetric legal objects associated with infrastructure developments. Therefore, legal data elements would need to be considered in the future extensions of open BIM standards for infrastructure. Currently, in the IFC alignment project, the concept of a spatial zone is considered for modeling boundary types associated with infrastructure assets. These boundaries include:

- Political boundaries: These boundaries define administrative regions such as states, counties, and cities. They play an underpinning role in specifying governing authorities, legal permits, and legislation applicable to large infrastructure projects which may span several administrative regions spatially.
- Legal boundaries: These boundary types for multi-story buildings are explained in Chapter 5. Similarly, they can be defined in the subdivision and registration of stratified infrastructure assets.
- Usage boundaries: These boundary types specify planning and usage zones such as rights of way for future roads, drainage districts, or special assessment districts.

As described in this book, legal information is broader than the legal boundaries. Therefore, other legal data elements, such as spatial structures of legal spaces and their attributes, would need to be considered in managing the legally defined extent of infrastructure assets. In future, a BIM-driven approach will provide an integrated 3D digital space for managing all legal arrangements vertically stratified in aboveground and underground built assets.

7.4 Spatial Reasoning of BIM Models

BIM tools provide the ability to define spatial and semantic properties of building elements for design and engineering purposes. These tools provide limited spatial relationships between two 3D objects, which results in limited queries of 3D data within the BIM environment. For instance, if a BIM model is exported to the IFC format, the IFC file may not contain some spatial relationships, such as *IfcRelContainedInSpatialStructure,* which are defined in the IFC schema. Additionally, spatial relationships between two 3D spatial objects are much more than the relationships considered in the current IFC standard. In other words, BIM models do not explicitly specify all the possible spatial relationships between two spatial objects. For instance, the "above" or "below" proximity relationships between a slab object and a wall object are not defined. Therefore, BIM tools should support advanced spatial reasoning methods to analyze and query 3D spatial datasets within BIM models. Spatial reasoning methods could be useful for retrieving the required 3D data and making spatial decisions in complex urban developments. BIM models for complex developments include a large number of spatial entities to store information describing physically existent elements as well as invisible space elements. Retrieving the required subset of information from such a complex and abundant data environment for a specific purpose is a very challenging task. In response to this change, 3D spatial query languages have been introduced to automate the extraction of essential spatial information from 3D BIM models (Daum and Borrmann 2014; Borrmann, Schraufstetter and Rank 2009). These languages rely on spatial operators which play an important role in handling large volumes of spatial datasets. Spatial operators are categorized into four types, namely:

- **Directional operators:** A directional operator is a binary relationship between two 3D spatial objects. Directions can be defined by either relative or absolute reference frames. In a relative reference frame, common relative directions between two spatial objects are *Back, Front, Left, Right, Below,* and *Above.* In an absolute reference frame, common directional operators are *West, East, North, South, Below,* and *Above.*
- **Topological operators:** Topological operators can be used to query the topological relationships between two spatial objects. The 9-Intersection model is the renowned topological formalism in spatial information science. A wide range of topological relationships between any two spatial objects are defined based on this model. Examples of topological relationships are *Contains, Overlaps, Touches,* and *Disjoints.*
- **Distance (Proximity) operators:** These operators describe the distance between two spatial objects. Examples are *MinDistance, MaxDistance, IsCloser,* and *IsFarther.*
- **Boolean operators:** These operators are used to perform logical operations between two spatial objects. Examples are *Union, Difference,* and *Intersection.*

As mentioned in Chapter 4, the boundaries of legal spaces are often determined by defining spatial relationships between these spaces and physical elements. For building boundaries, there are three common geometric connections of a legal space to its bounding building elements: (1) The legal boundary touches the interior face of a building element; (2) The legal boundary touches the exterior face of a building element; (3) The legal boundary passes through the median of a building element. In this book, these legal boundary types are modeled by defining semantic relationships between legal spaces and building elements. However, if it is assumed that semantic relationships between them are not defined at the stage of constructing a BIM model, topological operators can be applied to the geometry of legal spaces and building elements to automatically retrieve the spatial location of a building boundary. For any spatial object (A), the 9-Intersection model, proposed by Egenhofer and Herring (1990), subdivides a space with any dimension into three topological regions: (1) Interior of the spatial object (A°); (2) Boundary of the spatial object (∂A); (3) Exterior of the spatial object (A^{e}). According to this subdivision, the following 3 × 3 matrix is constituted to determine topological relationships between two spatial objects (A, B):

$$I = \begin{bmatrix} A^{o} \cap B^{o} & A^{o} \cap \partial B & A^{o} \cap B^{e} \\ \partial A \cap B^{o} & \partial A \cap \partial B & \partial A \cap B^{e} \\ A^{e} \cap B^{o} & A^{e} \cap \partial B & A^{e} \cap B^{e} \end{bmatrix} \tag{7.1}$$

The values of this matrix can be empty ($\varnothing$) or non-empty ($\neg\varnothing$). If it is assumed that both A and B are solids, then the possible topological relationships between two solids are: Disjoint, Contains, Inside, Equals, Touches, Covers, Covered by, and Overlaps. These relationships and their corresponding intersection matrix are represented in Figure 7.1.

Among the topological relationships, *Touches*, *Overlaps*, and *Covers* are relevant configurations for specifying types of building boundaries. If A is a legal space and B is a building element, the following relationships apply:

$$A \ \textit{Touches} \ B \rightarrow \textit{Interior Boundary} \tag{7.2}$$

$$A \ \textit{Overlaps} \ B \rightarrow \textit{Median Boundary} \tag{7.3}$$

$$A \ \textit{Covers} \ B \rightarrow \textit{Exterior Boundary} \tag{7.4}$$

Spatial analysis of 3D legal boundaries will support a wide range of spatial queries for vertically located properties. This includes queries such as:

- Where are the 3D legal boundaries associated with a private property?
- What are the legal rights associated with a building element?
- What is the association of a building element with its surrounding legal spaces?

Legal boundaries of 3D properties often cannot be easily perceived, due to the existence of complex structural and architectural elements inside

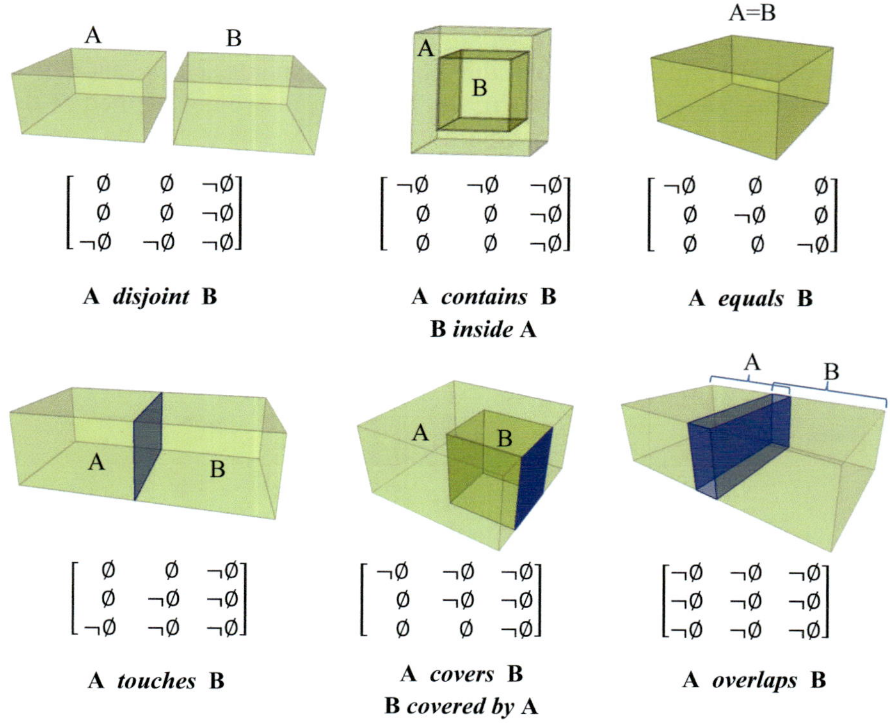

FIGURE 7.1
Topological relationships between two volumetric objects.

complex urban developments. Therefore, a 3D spatial query language for the BIM environment is required to automate the retrieval of various building boundaries associated with each legal space. This will help interest holders to easily identify legal boundaries intertwined with complex physical elements.

In addition to 3D boundary analysis, the development and implementation of a 3D spatial query language will bring new implications for a BIM-based land administration in complex urban built environments. It would provide the ability to leverage legal information for new urban applications, such as estimating the density of occupancy in 3D space (Rajabifard, Atazadeh, and Kalantari 2018). In addition, it will support 3D spatial analysis based on legal spaces which can be used for valuation and taxation purposes in complex developments. Another application is to perform a 3D buffer analysis on the legal spaces to identify underground and aboveground assets which are affected by a new development. Therefore, 3D spatial data analytics in the BIM environment would provide valuable intelligence in making spatial decisions associated with the legal ownership of vertically located properties.

7.5 Spatial Integrity of BIM Models

In this book, BIM models are constructed to manage volumetric legal spaces. However, these spaces may contain geometrical errors and may not follow spatial rules required for urban land administration. Therefore, it is required to check if the constructed BIM models include valid volumetric legal spaces. Spatial integrity refers to the rigorous procedures for checking, detecting, and rectifying possible errors in the legal partitioning of a BIM model. These procedures rely on a predefined set of validation rules. Validation rules are typically applied to an individual 3D legal space or a set of 3D legal spaces (Zhao et al. 2012).

The first spatial integrity check is to ensure that each legal space is defined by a spatially valid volume. In a BIM environment, solid modeling approaches are commonly used for defining the geometry of 3D spatial objects. However, solid models should be validated against the rules used for ensuring the spatial integrity of a single 3D legal space. To validate individual 3D legal spaces, one solution is to adopt the boundary representation (B-rep) and develop axioms for engineering validation rules accordingly. These rules will specify if the constructed 3D legal spaces are mathematically valid volumetric objects. One well-known method for confirming the volumetric closure of the 3D legal spaces is the Eular–Poincaré equation. B-rep-based solid models are defined based on a set of connected boundary surfaces. These boundary surfaces or faces are formed by a graph of edges and vertices. The representation of a B-rep-based solid model is shown by at least one shell. One shell, the outer, shall completely contain all the other shells and no other shell may contain a shell. As mentioned in Chapter 3, the Eular–Poincaré formula defines the following quantitative relationship among number of faces (F), edges (E), vertices (V), faces' inner loops (L), shells (S), and genus of shells (G) in B-rep solid models (see Equation (3.1)). In addition, Thompson and Van Oosterom (2011) proposed geometric validation rules specifically for 3D legal spaces, which can be also applied in the BIM environment. These rules are:

- The distance between two vertices should be less than ε.
- Each vertex must have at least three incident faces.
- The incident faces of a node must not intersect with each other. They can intersect each other at an edge.
- There must be at least ε distance between two non-intersecting edges.
- Every directed-edge of a face in the shell must belong to a fold. A fold is a pair of faces meeting at anti-equal directed edges with no other faces between them.
- The directed-edges defining a hole in a face must be part of the other faces.

- Bounded faces must be planar within the tolerance of ε.
- A node should have ε distance with a face, except if it is a constituent node of the face.
- No directed-edge may intersect a face except at a node of that edge.

The second level of spatial integrity is to check the validity of a legal interest associated with a set of legal spaces and physical elements. For instance, in the BIM environment, 3D digital representation of a common property is defined by composing various legal spaces and physical objects. Appropriate validation rules should be developed to ensure that the whole spatial structure of the common property is logically valid.

Finally, the connectivity relationship of each 3D legal space with its neighboring 3D legal spaces should be checked throughout the BIM model. In this context, validation rules for detecting gaps and overlaps between two legal spaces should be developed. Topological relationships could be used for the appropriate definition of the connectivity of legal spaces in 3D BIM models. For example, the "overlap" topological relationship can be used to identify colliding 3D legal spaces. This relationship is particularly useful for identifying those secondary interests affecting a primary parcel. On the other hand, the "touch" relationship can be used to identify those legal spaces which are topologically connected to each other in the 3D digital environment of BIM. This topological relationship plays a key role in avoiding gaps and overlaps between any two primary parcels in a BIM model.

7.6 Extending the IFC Standard to Other Jurisdictions

The adopted approach for extending the IFC standard, as described in Chapter 4, provides an underpinning framework for developing a BIM-driven urban land administration system in any jurisdiction, since it does not affect the current data structure of the IFC standard and incorporates legal data elements with as minimal a change as possible in the IFC standard, while also complying with built-in extension mechanisms in the IFC schema. Therefore, other jurisdictions can similarly develop an extension of the IFC standard for managing legal information in the 3D digital data environment of BIM in line with data requirements specific to the jurisdiction.

Appropriate identification of data requirements plays a key role in developing an extension of the IFC standard for urban land administration purposes in any jurisdiction. These requirements should be elicited by exploring the current urban land administration practices. Although the current practice in most jurisdictions are heavily reliant on 2D-based and fragmented representation mechanisms, the purpose is to mark out the 3D spatial extent of legal ownership for vertically located properties. The data requirements should be based on the important elements considered in the legal partitioning of an urban built environment within a particular jurisdiction. These elements can include, but are not limited to:

- Legal interests: A jurisdiction considers a wide range of legal interests for subdividing urban developments. Both drawn and notated objects defining the spatial extent of each legal interest should be identified. In some jurisdictions, physical objects are possibly considered in defining arrangements of legal interests. Therefore, it is important to understand how the arrangements and boundaries of each legal object are defined. Identifying attributes to be assigned to each legal interest is also very important.
- Legal boundaries: Each jurisdiction defines its unique legal boundaries which are specific types of the well-known general and fixed boundaries. An understanding of the boundary delineation methods provides the basis for identifying virtual denotations and physical objects defining legal boundaries.
- Land administration actors: In any jurisdiction, land administration actors play a wide range of roles and perform different tasks in urban land administration practices. Therefore, the key actors and their roles throughout the lifecycle of an urban development should be identified.
- Legal documents: Each jurisdiction has its own legislation for issuing and exchanging legal documents in urban developments. These documents and their important attributes are vital during the various stages of land administration processes in urban areas.

After eliciting data requirements in a particular jurisdiction, the suggested IFC extension approach can be used to enable a BIM-driven urban land administration within that jurisdiction. IfcZone and IfcSpatialZone can be considered as appropriate entities to define spatial arrangements of legal interests. If a legal boundary is defined based on measurements in a jurisdiction, the IfcVirtualElement entity should only be used to model this legal boundary inside the IFC standard. The concept of property set in IFC provides a good mechanism for incorporating attributes of each legal interest without significant changes in the current data structure of the IFC standard. For instance, in this book, the "Pset_Lot" property set was used to incorporate legal attributes of a "lot" to the IFC standard. Likewise, other jurisdictions and countries can develop their own property sets for meshing legal attributes into the IFC standard. Another important concept is user-defined values which can be used for specializing general concepts for land administration purposes. For example, a spatial zone is a general concept for aggregating physical and space elements. However, its type can be specialized into a specific legal zone by assigning the name of the legal interest. In this book, it was shown how various legal zones in the Victorian jurisdiction can be specialized from the general concept of IfcSpatialZone or IfcZone. Similarly, other jurisdictions can develop their own specialized legal zones in the IFC standard.

7.7 BIM for Property Measurement in Urban Land

Measuring the area and volume of properties plays a significant role in determining the value of properties. Therefore, property measurement is important for urban land management. There are several international property measurement standards which specify the area and volume of buildings and building parts, such as performance standards in buildings (ISO9836 2017), the International Property Measurement Standards (IPMS), the Royal Institution of Chartered Surveyors (RICS) code of measuring practice, and the Property Council of Australia (PCA) methods of measurement. The types of volumes and areas specified in these property measurement standards can be modeled by using one or more instances of the "IfcSpace," which is the entity used for modeling any type of space inside buildings. The boundary types identified in this book were specifically used for modeling legal spaces. However, they can similarly be used for delineating the spatial extent of functional areas or spaces defined in property measurement standards. For instance, in the RICS code of measuring practice, Gross Internal Area (GIA) is considered as the area of a building measured to the internal face of the perimeter walls at each floor. In this case, the interior wall boundary type modeled in a BIM environment can help communication of the GIA area and its correct computation. There are currently some quantity sets in the IFC standard, but there still remains a limited investigation into the subtleties with respect to the appropriate modeling of property measurement boundaries and computation of those quantity sets within the BIM environment.

A potential future direction for a BIM-driven urban land administration is to investigate the role of a BIM environment and the IFC standard in facilitating communication of boundaries for area and volume types used in property measurements. This will explore the viability of the connection between property measurement standards and BIM, which is very likely to provide a digital, interactive, and semantically optimized information environment to support the property measurement and integrated management of buildings during their entire lifecycle. Therefore, using a BIM-driven approach will provide the ability to link property measurement information with the building lifecycle in a 3D digital environment. On the one hand, a BIM-driven approach will reduce the reliance on 2D-based current property measurement practices in complex buildings, which have limitations in the communication of various boundary types in complex property spaces in buildings. On the other hand, it will be beneficial for further development of BIM, and more significantly, will establish a digital information platform for property measurement processes, which will subsequently optimize the management of property information in complex urban settings.

Compared to the traditional 2D-based property measurement approaches that use a fragmented method to compute areal and volumetric quantities, the 3D digital environment of BIM provides an integrated and automated method to compute these quantities.

The interactivity of a BIM environment improves the tasks associated with the visual inspection and checking required for a property measurement. A BIM model can be viewed in different modes including representations of area, volume, and boundaries of properties. The productivity would also be improved in a BIM environment, since it could automatically calculate all the values of the areas and volumes of the properties, combine and manage all the relevant documentations associated with property measurements within a single model profile, and any modification applied in any parts of the project within the BIM model will be coordinated by the model itself, which reduce the time and cost of the labor force to a large extent. In addition, the current standards do not provide specific guidelines for the property measurement in spatially complex structures. The representation of example property boundaries using BIM can be helpful for providing particular guidelines when measuring properties in buildings with a complex and irregular shape.

7.8 Automatic Generalization of BIM Models for Land Administration Purposes

BIM models provide highly detailed spatial and semantic information describing building components. The detailed representation of building components is not required in urban land administration practices. In some cases, very detailed information may add to the complexity of the BIM model, which may hinder the interaction of users with the BIM environment. Information redundancy in BIM models could be a technical issue for urban land administration practices. Not all the architectural and structural building elements are used in modeling legal arrangements and boundaries of urban properties. Therefore, inessential building information, which is not important in the context of urban land administration, should be removed from BIM models. A good generalization technique would provide a suitable way to reduce redundant building elements to simplify spatial and semantic complexities in the BIM environment. Development of these generalization approaches will automate extraction of required building elements from BIM models and eliminate unnecessary elements. Considering that A is a legal space and B is a building element, the containment topological relationship can be used as one potential solution to eliminate inessential elements from BIM models for the purpose of urban land administration:

$$A\ Contains\ B \rightarrow Eliminate\ B\ from\ the\ model \quad (7.5)$$

References

Borrmann, André, Stefanie Schraufstetter, and Ernst Rank. 2009. "Implementing Metric Operators of a Spatial Query Language for 3D Building Models: Octree and B-Rep Approaches." *Journal of Computing in Civil Engineering* 23 (1):34–46.

Daum, Simon and André Borrmann. 2014. "Processing of Topological BIM Queries Using Boundary Representation Based Methods." *Advanced Engineering Informatics* 28 (4):272–286.

Egenhofer, Max J. and John Herring. 1990. "Categorizing Binary Topological Relations between Regions, Lines, and Points in Geographic Databases." Vol. 9.

ISO9836. 2017. *"Performance Standards in Building – Definition and Calculation of Area and Space Indicators."* Geneva, Switzerland: International Organization for Standardization.

Rajabifard, Abbas, Behnam Atazadeh, and Mohsen Kalantari. 2018. "A New Method for Integrating 3D Spatial Information about Vertically Stratified Ownership Properties into the Property Map Base." In: *FIG Congress 2018 Embracing Our Smart World Where the Continents Connect: Enhancing the Geospatial Maturity of Societies*, 6–11 May 2018, Istanbul, Turkey. International Federation of Surveyors (FIG).

Thompson, R. J. and P. J. M. Van Oosterom. 2011. "Axiomatic Definition of Valid 3D Parcels, Potentially in a Space Partition." In: *Proceedings 2nd International Workshop on 3D Cadastres*, 16–18 November, 2011. Delft, the Netherlands. International Federation of Surveyors (FIG).

Zhao, Zhigang, Renzhong Guo, Lin Li, and Shen Ying. 2012. "Topological Relationship Identification in 3D Cadastre." In: *Proceedings of the 3rd International Workshop on 3D Cadastres: Developments and Practices*, 25–26 October 2012, Shenzhen, China.

Appendix A: EXPRESS-G Notation

IFC standard is primarily developed based on the EXPRESS language. To facilitate understanding of the proposed entities and their relationships, EXPRESS-G notation, which provides a graphical representation of EXPRESS lexical entities, was used in this book. The main EXPRESS concepts with their graphical notation are described below:

- Entity: An entity defines a class for objects having similar characteristics or behavior. In EXPRESS-G, an entity is represented by a rectangular box, and the name of the entity is written inside the box. In the IFC standard, entity boxes are highlighted in yellow (see Figure A.1).
- Attribute: Attributes are used to describe the characteristics of an entity. Attributes can be defined as simple data types, such as integer, string, real, or Boolean, or complex data types such as enumeration, selection, or referencing other entities. In EXPRESS-G, the name of simple data type is written within a continuous box having two vertical lines at the right end of the box. Attributes referencing other entities can also be considered as relationships between two entities. In EXPRESS-G, an attribute is represented by a solid or dashed thin line which is terminated by a circular arrowhead. The attribute name is written alongside its corresponding line. A dashed line represents an optional attribute, while a continuous line is used to show a mandatory attribute. In the IFC standard, all data types are highlighted in green (see Figure A.1a).

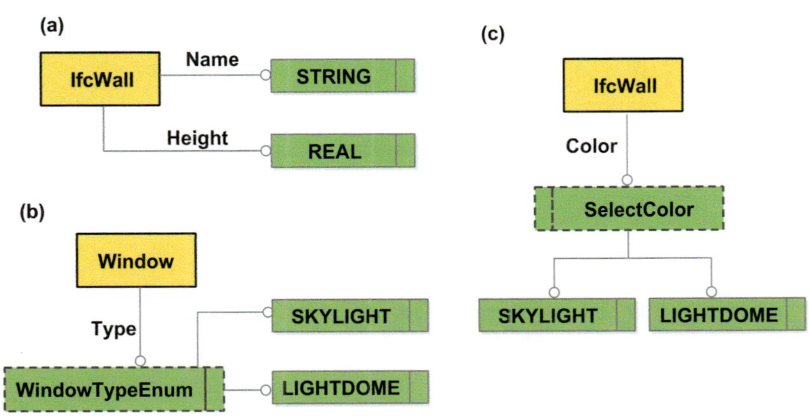

FIGURE A.1
Examples of attributes with different data types in EXPRESS-G notation: (a) Simple data types, (b) Enumeration, (c) Selection.

- Enumeration: An enumeration data type refers to a range of values belonging to its corresponding attribute. Only one of the enumerated values can be used in an instance of the entity. In EXPRESS-G, the name of an enumeration data type is written within a dashed box having two vertical lines at the right end of the box (see Figure A.1b).
- Selection: A selection data type indicates that exactly one of the alternatives must be selected in an instance of the entity. In EXPRESS-G, the name of a selection data type is written within a dashed box having two vertical lines at the left end of the box (see Figure A.1c).
- Relationship: A relationship expresses a dependency or interaction between two entities. In EXPRESS-G, relationships are represented by a solid or dashed thin line which is terminated by a circular arrowhead. The relationship name is written alongside its corresponding line.
- Cardinality: A relationship has cardinality, which indicates the number of objects in each of the entities at either end of the relationship that may be involved in a specific instance of that relationship. Cardinalities can be expressed in terms of sets or lists. A set is an unordered, variable length collection of unique items. A list is an ordered, variable length collection of not necessarily unique items; to a list of unique items. In EXPRESS-G, A cardinality can be shown as a string in the form C[m:n], where C is one of S(set) or L(list), m is the minimum number of items allowed in any instance of the relationship, and where n is the maximum number of items allowed in any instance of the relationship (see Figure A.2a).
- Inverse relationship: In most cases, an inverse relationship can be inferred directly from the original relationship. As shown in

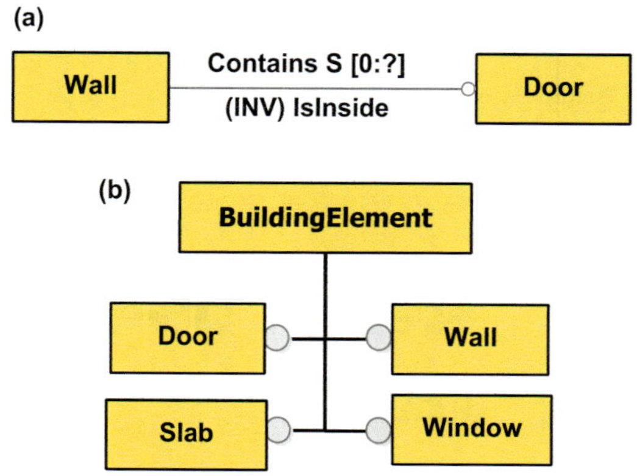

FIGURE A.2
(a) Example of relationships between two entities; (b) Example of inheritance.

Figure A.2a, an inverse relationship is indicated by writing (INV) before the name of the relationship.

- Generalization and specialization of entities: Two or more entities which have some characteristics or behavior in common may be generalized into a supertype. In other words, each subtype entity is a specific specialization of the supertype entity, inheriting all characteristics and behavior of the supertype; however, each of them has also its own distinct characteristics or behavior. In EXPRESS-G, the relationship between subtypes and supertypes is shown by thick continuous lines with circles touching subtype entities (see Figure A.2b).

Appendix B: Business Process Modeling Notation (BPMN)

Business Process Modeling Notation (BPMN) is a method of illustrating business processes in the form of a diagram. This Appendix provides an informative description of those BPMN symbols used in this book (see Figure B.1). These symbols include:

- Activity (Task): A task is an atomic activity that is included within a process. A task is used when the work in the process is not broken down to a finer level of process detail.
- Start Event: A start event indicates where a process will start.
- End Event: An end event indicates where a path of a process will end.
- Message: A message symbol can be used to either send or receive a message.
- Timer: It indicates the maximum time in which an activity should be completed.
- Exclusive Gateway: It routes the sequence flow to exactly one of the outgoing branches.
- Data Object: Data objects provide information about what activities require to be performed and/or what data they produce. Data object elements must be contained within process or sub-process elements.
- Sequence Flow: It is used to show the order in which activities will be performed in a process and in a choreography.
- Message Flow: A message flow is used to show the flow of messages between two participants that are prepared to send and receive them.

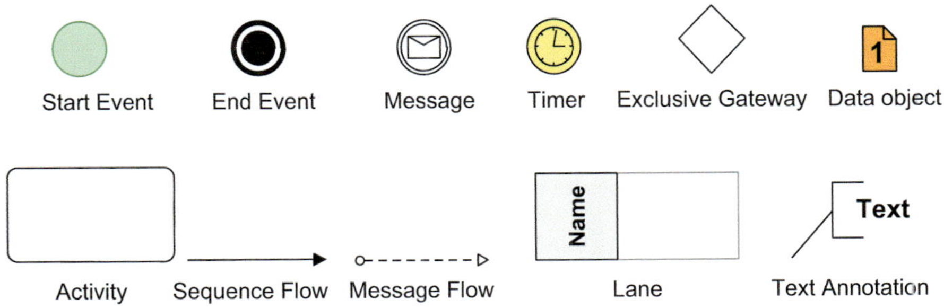

FIGURE B.1
A subset of BPMN symbols used in this book.

A message flow must connect two separate pools. They connect either to the pool boundary or to flow objects within the pool boundary. They must not connect two objects within the same pool.

- Lane: A lane is the graphical representation of an actor in a collaborative process.
- Text Annotation: Text annotations provide additional information for the reader of a BPMN diagram.

Index

For Product Safety Concerns and Information please contact our EU representative GPSR@taylorandfrancis.com Taylor & Francis Verlag GmbH, Kaufingerstraße 24, 80331 München, Germany